BOSTON STUDIES IN THE PHILOSOPHY OF SCIENCE

VOLUME VI

ERNST MACH: PHYSICIST AND PHILOSOPHER

SYNTHESE LIBRARY

MONOGRAPHS ON EPISTEMOLOGY,

LOGIC, METHODOLOGY, PHILOSOPHY OF SCIENCE,

SOCIOLOGY OF SCIENCE AND OF KNOWLEDGE,

AND ON THE MATHEMATICAL METHODS OF

SOCIAL AND BEHAVIORAL SCIENCES

BOSTON STUDIES IN THE PHILOSOPHY OF SCIENCE

VOLUME VI

ERNST MACH
PHYSICIST AND PHILOSOPHER

EDITED BY ROBERT S. COHEN AND RAYMOND J. SEEGER

D. REIDEL PUBLISHING COMPANY / DORDRECHT-HOLLAND

Printed in The Netherlands by D. Reidel, Dordrecht

ERNST MACH

PREFACE

At the annual meeting of the American Association for the Advancement of Science in Washington, D.C., 27 December 1966, a symposium was held to commemorate the 50th anniversary of the death of Ernst Mach, the physicist who was vitally concerned about philosophical foundations. It was arranged by Section B on Physics, and co-sponsored by Section L on the History and Philosophy of Science, as well as by the History of Science Society. Dr. Allen W. Astin, Vice-President of the Association and Director of the National Bureau of Standards, presided. Representing the Austrian ambassador, Dr. Ernst Lemberger, a few opening remarks on his behalf were made by Dr. Walter Hietsch. Also present was Dr. Ernest A. Lederer, a grandson of Ernst Mach.

The contributors, to the symposium, mostly physicists, represented different backgrounds and differing points of view; they presented their review of Mach's work primarily in the light of subsequent developments. They all, however, share a common interest in the life and works of Ernst Mach. Two of them, Otto Blüh and Peter G. Bergmann, received their doctoral degrees in theoretical physics from the University of Prague. Karl Menger received his doctoral degree in mathematics from the University of Vienna (he is responsible for the latest edition [1960] of Mach's celebrated *The Science of Mechanics: A Critical and Historical Account of its Development*, for which he prepared a new Introduction). Robert S. Cohen and Raymond J. Seeger received their doctoral degrees in theoretical physics from Yale University where each began a continuing interest in the development and philosophical foundations of scientific thought. Floyd Ratliff, who received a doctorate in psychology from Brown University, is a neurophysiologist, and recently (1965) wrote a definitive work on *Mach Bands: Quantitative Studies on the Neural Networks in the Retina*. The only professional historian of science in the symposium, Erwin N. Hiebert, received his doctor's degree from the University of Chicago with particular emphasis upon the history of chemistry since 1800. Each of these speakers gave a 30-minute presentation. The present articles are amplifications of their remarks on that occasion. Some duplication of content among them have not been deleted. In addition, this volume contains papers by Wolfgang F. Merzkirch, Director of the Ernst-Mach-Institut in Freiburg, Germany, by Gerald J. Holton of the Department of Physics at Harvard University, and by Hubert Goenner of the Institute for Theoretical Physics at the University of Göttingen. (Dr. Merzkirch had been invited to participate in the symposium, but was unable to do so. Dr. Ratliff, too, was not present, owing to illness.) The reader will note that the interpretation of Mach's life and works is still quite unsettled. The editors have made no attempt to reconcile differences of opinions expressed by the several participants, or even to eliminate overlapping comments, which are often from different points of view.

One of us (R.S.C.) suggested that the address given by the late Richard von Mises on the occasion of the centennial of Mach's birth be included. Although published, it is inaccessible in English. A translation of this document was made possible by a grant from the Institute of the Unity of Science; it is included in the Appendix. In view of Mach's decisive influence on the Vienna Circle, it was then decided also to include two commemorative addresses by our friend and teacher the late Philipp Frank, one on the occasion of Mach's death and the other at the centennial of his birth. For convenience of readers, the Appendix also contains material on the life and publications of and about Mach, compiled primarily by Otto Blüh.

The Boston Colloquium for the Philosophy of Science is one of many groups and institutions which have been inspired by the combination of conceptual analysis, empirical relevance, critical rigor and humane motivation which was Ernst Mach's. Through the Vienna Circle's *Ernst Mach Verein*, and then the Institute for the Unity of Science, which took up its second home in Boston and Cambridge, and through the sponsorship received by Philipp Frank, the Boston Colloquium is linked with Mach. It is therefore appropriate and pleasant to publish these expanded studies of Ernst Mach as a volume of *Boston Studies in the Philosophy of Science*.

ROBERT S. COHEN
Boston University

RAYMOND J. SEEGER
U.S. National Science Foundation

TABLE OF CONTENTS

OTTO BLÜH

ERNST MACH – HIS LIFE AS A TEACHER AND THINKER

I

The first recent occasion for commemorating the work of Ernst Mach presented itself in 1938, the year of Mach's hundredth birthday. At that time the difficult political situation in Central Europe made any such event unfeasible. The members of the so-called Vienna School of neo-positivism, who at their beginning had assumed the name of 'Ernst Mach Society', had dispersed, and in Prague, where a program of lectures on Mach had been prepared, the celebration had to be curtailed due to the unrest in the spring of 1938. The spring of 1966 offered the next opportunity for remembering Mach and his work, on the fiftieth anniversary of his death in 1916. In fact, the Ernst-Mach-Institut in Freiburg i. Br. arranged an international conference on Mach, and we here, with this symposium, will be trying to do our part in honoring Mach's personality and his work.

The postponement of 28 years has had its definite advantages. In 1938 any celebration would have been a partisan affair and would probably have done more harm than good to the lasting memory of Mach. In the meantime we have gained a much better perspective and can appreciate Mach's work in its full extent. We should, today, be able to discuss his work objectively and even be prepared to admit its shortcomings. Our symposium, I believe, does not represent any definite scientific or philosophic opinions, and this meeting is in no way a partisan 'Teach-in'. It should help us to get acquainted with the work of a scientist of whom even the scientific world knows little. Mach's name has become familiar through the Mach number, but most scientists connect Mach only vaguely with definite scientific achievements or philosophical views, and not always correctly. His role as a historian of science has been acknowledged and challenged. Least known is Mach's work as a teacher, work which has been taken for granted. My task in this symposium is in particular to show you that part of Mach's work in which he took an abiding

Boston Studies in the Philosophy of Science, VI.

interest, and to bring it into context of his general studies and ideas. My thesis is that Mach's thinking arose primarily from a need for didactic clarification and led him to search for the limits of man's intellectual capacities to understand himself and the world.

The external circumstances of Mach's life can be told in a few words. Biographical information is now readily available in English, thanks to the recent book on *Mach Bands* by Dr. Floyd Ratliff[1]; and there is also a modern German-language account of Mach's life by the late Dr. K. D. Heller, of Haifa, which contains a cross-section from Mach's writings.[2] Mach was born on February 18, 1838 in Turas (Tuřany), in the province of Moravia, then part of Austria, now Czechoslovakia, a province from which had come Amos Comenius, Gregor Mendel and Sigmund Freud. The Mach family moved into the neighbourhood of Vienna when Ernst was a young child, and many years were spent in the country where educational facilities were scanty and Mach's education depended greatly on tuition by his father who had been a high-school teacher and whose ambition was to be a farmer. Mach began his university studies in Vienna in 1856 and finished them in 1860. Shortly afterwards, in 1861, he became a university docent of physics. In 1864 he was appointed professor of mathematics and physics at the University of Graz, and in 1876 made professor of physics and director of the department of physics at the University of Prague. Mach lived in Prague for 28 years, as a teacher, experimenter and scholar, active in university affairs. These were the most fruitful years of his life. Evidently he was not influenced by the mystique of the place, and nothing in his writings reminds us of the Kafka-esque anguish and anxiety. Frequent visits to Vienna helped to maintain contact with friends, in particular with the sociologist and writer Josef Popper. In 1895 Mach was called to Vienna to occupy a newly established chair of philosophy for the 'History and Theory of the Inductive Sciences', mainly through the good offices of the classical scholar Theodor Gomperz, who was also editor of the German translation of John Stuart Mill, and who wanted to strengthen liberal ideas in the faculty. It was only for a few years that Mach had opportunity to lecture on the history of science, since three years after his return to Vienna he suffered a stroke which paralyzed the right side of his body, fortunately without affecting his mental faculties. When in 1901 it became impossible for him to fulfill his teaching obligations he insisted on his retirement.[3] Mach's successor till 1905 was Ludwig Boltzmann. Mach lived in Vienna till 1912; then he moved to the home of his son, Dr. Ludwig Mach, near Munich, where he died on February 19, 1916. Ludwig Mach was a doctor of medicine, but he did not practice, and he was for many years his father's devoted assistant, and after his death became the trustee of Ernst Mach's papers.

Ernst Mach had the normal career of the pre-1914 European university professor; a life spent near home, devoted to teaching and research; with time for contemplation and recreation in the company of congenial friends. Mach experienced personal misfortunes and a sense of the tragic nature of life was not unknown to him from the days of his youth. He had the satisfaction of seeing his views shared by a small group at home and abroad. In America, which he never visited, his writings became known through the efforts of Dr. Paul Carus, and through publications in the magazines *The Monist* and *The Open Court*. Other Americans he had met personally, or had correspondence with, were William James, J. B. Stallo and Jacques Loeb. His long illness did not prevent him from reviewing and defending his thoughts, but probably proved a serious handicap for a revision and expansion of his views in the light of the developments in the sciences in the first decade of this century.

II

I shall now turn to a discussion of Mach's years as a student and young lecturer in Vienna and to a look at his intellectual development against the background of the academic locality and the general scientific situation in the sciences of this period. Mach's professors of physics were Andreas (Ritter) von Ettinghausen and August Kunzek; the former is known as the discoverer of a thermo-electric effect related to the Hall effect; he was the director of the physics department as the successor of Christian Doppler, who had occupied this chair for only two years. Ettinghausen was the first professor in a German university to introduce organized laboratory exercises and training in physical experimentation. Before, physics had been taught through demonstration experiments and required no more than a 'Kabinet', a room for the collection of demonstration apparatus for which Kunzek was responsible. Josef Stefan was only a few years older than Mach and was a young lecturer when Mach was a student; he became the successor of Ettinghausen in 1866. Josef Loschmidt, who began as a practical chemist and was much older than Stefan and Mach, became professor only in 1868 (to 1891) and lectured then – as one would expect – on molecular physics. Ludwig Boltzmann was a student at Vienna between 1861 and 1865 and may have attended Mach's first lectures. The luminaries of the physical sciences faculty were the mathematician Josef Petzval, who also lectured on oscillations of elastic bodies, and became famous for his design of the first achromatic double-objective, and the astronomer Carl von Littrow. (Littrow the younger; his father, Josef Johann Littrow, was the author of the popular astronomy book *Wonders of the Sky*.)[4]

Compared with the general situation of European physics at the time, the physics in Vienna in the eighteen-fifties and sixties was not in the forefront of modern research, a position which it attained in the eighties through the triumvirate Stefan, Loschmidt, and Boltzmann. The principle of conservation of energy made its way only slowly into the teaching of the universities, and thermodynamics and kinetic theory, connected with the names of Hermann von Helmholtz and Rudolph Clausius, had at Mach's time no representatives in Vienna. The same can be said about the field of electrodynamics of Wilhelm Weber and William Thomson (Kelvin). Mach's great friend Josef Popper studied at the polytechnic in Prague between 1854 and 1857 and, as he tells in his memoirs[5], heard about the principle of the conservation of energy for the first time in a lecture by another physics lecturer in Vienna, Ernst Reitlinger, in 1862, that is, twenty years after Mayer's and Joule's discoveries.

It is therefore not at all astonishing that Mach in his undergraduate years was attracted to the work of two great personalities of the biomedical sciences in Vienna, the physiologists Ernst Brücke at the university, and Karl Ludwig, at an affiliated medical academy and hospital. In modern jargon, it was there 'where the action was.' Brücke was particularly interested in vision and speech; Ludwig, in basic problems of physiology; he was the inventor of the kymograph for the recording of physiological phenomena. Both held leading positions, with Helmholtz, Emil Du-Bois Reymond, and Rudolf Virchow, in the anti-vitalistic movement and attempted to make biology into biophysics, that is, to make it into a causal life science. They used various physical methods in their physiological investigations and pointed to a new and wide field of applications of physics. Mach was impressed by this modern research activity, and his numerous publications on psychological and physiological problems between 1860 and 1875 indicate the strong influence of the physiologists of the Vienna school, which later provided the basic material for his scientific philosophy.

I want to mention here yet another possible influence on Mach, namely the great significance given in Mach's time to the pedagogical ideas of the philosopher and psychologist Johann Friedrich Herbart. Herbart, who at one time had held Kant's chair in Königsberg, shared his belief in a strict separation of theoretical and practical (i.e., moral) philosophy. His writings on psychology as a science, and on the possibility and necessity of applying mathematics to psychology, had their effect on many scientists and admittedly also on Mach (Ref. 21, p. 299). Herbart was later professor at Göttingen and never taught at Vienna – he died in 1841 – but his views became widely known in Austria through the influence of the philosopher Franz S. Exner, who from 1848 was an adviser to the minister of education and planner of the middle-schools or gymnasia in Austria, where the main ideas of Herbart's psychology of education were to be realized. I believe that the general ideas of Herbartian didactics left their mark on Mach, especially his stress on organization and presentation of the subject matter in accordance with the storage processes of the learner's mind, that is, an economics of learning.[6]

I thus recall the Viennese academic climate in order to show the background of Mach's early intellectual life. The totality of his philosophy or world view is, however, not entirely understandable from a knowledge of academic influences only. One has to understand the political, social, and religious atmosphere in Austria after 1848, with its struggles among clericalism, liberalism, nationalism, and individualism. But it is probably true to say that the freedom of Mach's thought was nurtured in the enlightened atmosphere of his family and strengthened through the opinions of congenial friends and collaborators during the years of maturity. A closer investigation of all these factors in Mach's life is still missing.

III

Mach's first attempts in teaching were not in pure physics, but in physics related to physiology. As a result of Mach's affiliation with the medical school he offered a lecture course in physics for medical students. In this course he impressed the clinical professors so much that he was nominated for a professorship in natural science at the then existing medical academy at Salzburg, an appointment he declined, and which would have side-tracked him into a subordinate position, with few prospects for advancement.

After one year's experience in teaching he published a *Compendium of Physics for Medical Students*, issued in 1863 in Vienna[7]; this was evidently a labor of love, intended as an educational message. Mach dedicated this book of about 270 pages and 225 illustrations to Ettinghausen – *captatio benevolentiae*, but there is no reference of support by an Austrian N.S.F., nor any note that this was a group project. In the Preface he explains that the book should serve as a 'Leitfaden' or 'guide-line' to his lecture-course, and that he had set himself the task "to acquaint his audience with the best accepted physical concepts in as short a time as possible". The last clause seems to be somewhat foreign to the tempo which one usually connects with the city on the blue Danube, and one must accept it as a sign of Mach's extraordinary, youthful impatience; it also somehow anticipates Mach's later emphasis on 'economy' in science. He recognized quite clearly that "what makes the study of physics difficult for the student of medicine is not the ignorance of detail, but rather the unfamiliarity with the physical conceptual modes in general" (p. V). He explains that there is a growing need for breaking down the compartmentalization of physics, since "the now existing system is becoming, day by day, more impractical" (p. V). It is most interesting in view of Mach's later standpoint that he tried to develop a unitary theory of physics with the help of the atomic theory, or as we would say now, the molecular theory, which he applied to explain a variety of physiological phenomena. A considerable part of the book is devoted to discussions of blood circulation, the mechanics of the heart, ear and sound, the source of animal heat, heat economy of animals, light sensation, and color, etc.

Regarding the atomic hypothesis, the preface to the *Compendium* already sounds a warning, probably inserted after the course had been given and experience gained with this approach. Mach says here that atomism establishes nothing more than simple models and that the theory itself would need further explanation and is not "the last and highest" (p. VI), and further: "One can also consider the atomic theory as a formula which led to many results and in future may lead to more. In fact, whatever will be the future metaphysical view of matter, it should be possible to translate the results of the atomic theory into it, as one can express formulas in polar-coordinates or in parallel-coordinates" (p. VI). This is certainly a dualistic view of physical nature and must be regarded as one of Mach's many intuitions of things to come. A few years later Mach suggested that one could consider "molecular events musically", that is, as vibrating systems (Ref. 18, p. 51); another *aperçu* in truly oracular formulation.

Compared with the usual physics texts of the time, apart from the biophysical bias, Mach's *Compendium* is distinguished by its philosophical intention. It omits the detailed description of physical apparatus, or as Mach says in the Preface (p. VI), "of finely polished little hooks and screws", as they could be found in the elaborate woodcuts, for example, in Müller-Pouillet's general physics text, or in the similar French works of Jamin and Ganot, from which they found their way into many physics texts still in use at the beginning of this century.

In agreement with his stated intellectual aims, Mach makes it clear (p. 1) that most physiologists at the time believed that "the phenomena in animal bodies are essentially the same as in inorganic nature; that the vital phenomena are brought about only through the cooperation of physical forces, that one therefore could dispense with the concept of a vital force, which, if one wants to retain the name, is only a general term for the collection of the physical forces acting in the organism". He then discusses the connection between the hypotheses of vital force and of teleology. He considers the assumption of a purposeful activity of living nature as a mistake "since the concept of an aim presupposes a personal interest which forms itself in the life of man and also could only be applied to it.... The carpenter obtains his concept of aim from the joiner's bank, the smith from the anvil and it is evident that such an arbitrary transfer of this concept to the events in nature is a *coup de force* (*Gewaltstreich*)". Mach quite often uses a very plastic language in expressing his convictions, though he never neglects in any argument also to show the other side of an argument. Thus he does not deny (p. 3) that a teleological standpoint could be used as an instrument of research: "Animals live under certain conditions, we therefore can ask the question: 'Through which arrangement is the existence of animals under the given condition made possible?'".

Astonishingly, Mach does not mention Brücke, Ludwig or other anti-vitalistic German biologists of the time, nor the French school of experimental physiologists, for example, Claude Bernard, whose *Experimental Medicine*, however, was published only in 1865; but he clearly identifies the scientific study of physiology with a study of physics. "It does not need a deep penetration to recognize the progress made", he writes, enumerating the fields of blood circulation, movement of muscles, endosmotic processes, etc. He also points out that physiology increasingly "falls under the sway [*fällt anheim*] of mathematics. Since the knowledge of physicians is mainly founded on physiology, an understanding of physics becomes an essential factor in the study of medicine". There follows information on the organization of the *Compendium*, in which he promises to look at the atomic theory which "so to speak forms the philosophic terminus [*Abschluss*] of physics". Discussing the methods of research he also points out (p. 7) the problem of evaluating random results or how to eliminate the effects of chance (*Elimination des Zufalls*), in particular of biological observations. "Life processes are determined through a series of conditions, of which we can only observe a small part, and to a still smaller degree have it in our power [to influence them]. For just this reason chance plays a great role and one cannot draw any certain conclusions from one or a few observations. The mathematical-statistical method is therefore for clinical observations to be highly recommended, and one can well-nigh state that it alone can lead to salvation". Mach writes '*zum Heile führe*', which can also mean 'can lead to a cure' ['*heilen*' = to cure] doubtless a purposeful, somewhat facetious ambiguity.

It would probably be of interest to both physical and biological scientists to review the contents of the *Compendium* in greater detail, and I may do this another time. I only want to mention one anecdotical example which shows Mach's sense of humor and, since it probably was born in the lecture room, discloses Mach's attitude at the lectern. The question had been raised how much heat is produced through sound. Mach, in a footnote (p. 170) says to this, "When Redtenbacher ... asserts that sound changes little into heat, since we cannot heat a concert hall with a fugue by Bach, maybe one could convince oneself of the opposite by listening to a composition by R. Wagner". In Vienna, where the famous music critic, and historian of music at the university, Eduard Hanslick had been the sharpest adversary of Richard Wagner's new musical style, the *ad lib* of Mach probably had great success in the classroom and the young docent could not resist the temptation to immortalize the joke in print. That Mach, however, gave his auditors also occasion for more serious thought may be exemplified by his reference to Wilhelm Weber's theories of electrodynamics (pp. 256–257), when he explained that the dependence of electrodynamic forces, for example, between electric currents, depended on the velocity and is "at present still a puzzle; it is not found anywhere else, where two bodies (e.g., through gravity) act directly on one another". It would seem that in this latter statement can be found the latent seed to the problem whose answer culminated in the renowned 'Mach Principle' of relativity.

The copy of the *Compendium* which I was privileged to use – lent to me by the Ernst Mach Institute[8] – was Mach's own desk copy, and was evidently used by him for many years in Prague, where the majority of his students in the main lecture course were first-year students of medicine. Into the book are bound blank pages on which Mach had made numerous remarks and bibliographical notes. That Mach did not attempt, however, to bring out a second edition, would seem to indicate that his efforts had not been very rewarding. Apparently, his medical students did not always appreciate his endeavors on their behalf. He was frequently questioned by medical students who wanted to hear from him "how much physics will a doctor need?" Mach's answer was "A doctor always needs exactly as much physics as he knows. Knowledge one does not possess one cannot use". The German original version makes use of the double-meaning of the word '*brauchen*', which means 'to need' as well as 'to use'.

IV

While lecturing in Prague, Mach developed other aspects of a physics teaching program. Mach was an experimental physicist, and many of his investigations in physics and psychology show his talent in the field of experimentation. Faced with the task of lecturing to large classes (that means, probably no more than 100 students in the traditional experimental course), he directed his attention to the development of demonstration experiments, particularly in mechanics and optics. The apparatus shown in the illustrations for his *Mechanics* were built by his shop assistant Hajek and formed in Prague a well-kept 'Mach Collection'. Some of the apparatus was of original design, like the pendulum apparatus, the wave-machine, the optical smokebox, and others. A list of Mach's lecture courses in Prague shows that he had not only held the main class in experimental physics every year, 5 hours a week, but offered from time to time special lecture courses on heat, light, physics teaching in high schools, etc.[9] In Graz and Prague Mach also gave a number of public lectures which were later made available in print and were translated into English[18]. The variety of topics shows an astonishingly wide field of interests. The subjects and their treatment are never trite or trivial. (Cf. the headings in the bibliography of English translations at the end of this symposium.) Some questions of quite modern importance are included; for example, when Mach in his talk on 'The forms of liquids' (1868) mentions that on the moon statues made of syrup would have semi-permanence (a worthy project for any international outer-space agency in case they can decide on the object of the monument); one lecture on the velocity of light (1867) comes close to suggesting the famous thought-experiment with which Einstein disproved the vulgar meaning of simultaneity. A lecture 'On Symmetry' (1871) is devoted to the discussion of mirror images, which Mach relates to the human apparatus of vision. To mention just one more, in the lecture "On Transformation and Adaptation in Scientific Thought", delivered in 1883 on the occasion of Mach entering on his second Rectorship at Prague University, he accepted, with a reference to Darwin's theory, that knowledge is a product of organic evolution. It is significant for the liberalism of these days that Mach could advance this idea at a solemn university occasion with less danger of an official reprimand or public criticism than Father Teilhard de Chardin, who proposed more or less the same idea in our days.

At this time Mach did not lecture specifically on the history of physics, but his special lectures on heat, optics, electrostatics, etc. probably treated the subjects in a historical and critical manner, a method which he later accepted in his three books often styled historical: on mechanics[10] (1883), heat[11] (1896), and optics[12] (posthumously, 1922). These well-known books, of which only the *Mechanics* and *Optics* are available in English, were not really books of the history of these branches of physics, but were specialized textbooks. For example, the German title of the *Mechanics* reads literally: 'Mechanics in its Development', which does not give the same mephasis as 'The Development of Mechanics', so that the English translator quite correctly rendered the title as 'The Science of Mechanics,...'. [The German 'Entwicklung' has been translated as 'Development' but it can also mean 'Elucidation', 'Deduction' and 'Derivation'.] The full titles of the two other books are *The Principles of Heat* and *The Principles of Physical Optics*; they all have sub-titles which stress that the treatment is historical and critical. *Prima facie* these are books on mechanics, heat, and light, with a chronological approach, and a strongly critical attitude. In the preface to the book on heat[11] Mach wrote in 1896: "This book, just as the *Mechanics*, is on the one hand the result and on the other hand the basis of my lectures. Many a reader [that means teachers] must have had the experience that, relating generally accepted viewpoints with a certain enthusiasm, he suddenly recognizes that the matter does no longer come from the heart. Quiet considerations afterwards usually lead to the discovery of logical discrepancies which, once admitted, become unbearable." And later: "One should not expect in this work... the results of archivaric research; the question is to find the *coherence* and the *growth of thoughts*, not to find interesting curiosities.... Where I speak about persons, they interest me only as intellectual and at the best as *ethical individualities,...*" (pp. VII/VIII).

Mach has been taken to task for not having been more careful in his historical research.[13] If he failed in historiography because he may have ignored certain interpretations, we should not be too hard on him. Mach wanted to write not as a historian of science, but as a critical physicist whose work arose from a didactic purpose. He did not write the *Mechanics* to set the history of mechanics right, but to set Newtonian mechanics right; not to discuss Newton's achievement in its historical meaning, but in its scientific meaning. He tried to deduce for himself, his students and his readers, the development of certain branches of physics along logical, not chronological lines, and in spite of certain shortcomings these are still today, perhaps in the absence of something more perfect, considered useful and interesting for the student of the history of science.

It was in Vienna that Mach, in his new professorship on the history and theory of the inductive sciences, found it possible to announce lectures on 'The Development of Mechanics and of the Mechanical Sciences' and on 'The History of the Science of Heat and of Energy'.[14] His first lectures on 'Psychology and Logic of Research', of 1895/96 formed the basis (cf. Ref. 3) for his last work, *Erkenntnis und Irrtum* (Cognition and Error), published first in 1905[15]. In the preface to this book Mach refused to be called a philosopher since he felt that "the land of transcendence remains closed to me.... I have tried not to introduce a new philosophy into science, but to remove from it an old and stale one, a tendency which has been taken amiss also by quite a number of scientists.... in the course of forty years, in laboratory and lecture hall, I had the opportunity to recognize the ways in which knowledge advances"; and it is to questions of this nature that he addressed himself in this book. The individual chapters include thoughts already expressed elsewhere, but with many additional reflections, as the chapters on Thought-experiments, on Space and Time, on Hypothesis, and so on. At the end of the volume is a summary of quotations of Mach's own earlier writings which could serve as the nucleus for a much needed 'Concordance' of Mach's ideas, and which would be very valuable for the author of any future comprehensive biography.

V

Out of his extensive classroom experience in Prague, Mach also prepared a number of general physics texts (around 1890) for different levels of education: university or college physics, and higher and lower physics for the middle schools (high schools). One of these books was intended for the use of Mach's students and fulfilled, as he says, a 'local task'.[16] An identical printing under a different title was to be used in the advanced grades of Austrian gymnasia. Two physics texts for the higher schools of Germany were based on Mach's original one, but adapted to the regulations in the schools of the German Empire by two high-school teachers (1893/94), and went through several editions (Cf. the list of textbooks at the end of this article). Lack of time prevents me from discussing the contents of some of these books in detail, but I shall say at least a few words about the educational intentions of the author, and I shall quote a few lines which seem to have some significance for our present educational debates.

Mach was not an educator *per se* and he did not write at length on educational problems[17]. In his physics textbooks he did not try specifically to advance his own ideas. The basic tenets of his philosophy, however, to which belong his rejection of dogmatism and speculation, and an appreciation of historical understanding and liberal interpretations, are expressed appropriately as they apply to education. For example, in the preface of the general physics textbook he wrote: "Concepts are not dogmatically introduced ... but developed from facts. Theories and hypotheses come to discussion only when there is a need felt for them. Not through mentioning many dates and names, but through our genetic exposition does the historical moment gain its significance ... ." Mach did not apply long-winded mathematical deductions which, he believed, "did not contribute to understanding, but make it more difficult to find an overall viewpoint".

In education "one has to be intent on saving mental work. This is reached by analyzing the facts ... into simpler, recurrent parts. These parts, which alone we have to remember, will become particularly familiar to us. The capability of analyzing a new fact in already known parts we call ... the explanation of the fact." To mention only one idea of general curiosity, Mach points out in regard to molecules and atoms that these concepts have to be considered as hypotheses which should be accepted only tentatively, but should not be taken as definite truths (*ausgemachte Wahrheiten*). He admits that the relationship between heat and work, for example, gains in evidence (*Anschaulichkeit*) if one envisages bodies to consist of small particles and considers all physical events as mechanical; however, one should not forget that these ideas (*Vorstellungen*) are hypotheses.

Mach shows in his textbooks a prudent insight into the mentality of the average student. The words 'excellence', 'creativity' and 'challenge' in our modern educational jargon obviously did not belong to his educational vocabulary. In a lecture on the relative merits of education in the classics and the sciences[18] he also emphasized the importance of realistic and moderate aims in educational practice. "One point of importance here is that young students should not be exposed to premature abstractions, but should be made acquainted with the subject matter ... [and] that the more general point of view is to be postponed ... . If our young men are not to enter the universities with blunted, impoverished minds, if they are not to lose their vital energies in the high schools ... great changes will have to be made ... . [Otherwise] they leave as poor creatures who have learned too much ... [and] what they have acquired is a spider's web of thoughts too weak to furnish sure supports, but complicated enough to produce confusion."

In his autobiographical references Mach did not particularly mention his interest in education, probably because teaching belonged to his everyday work. That he felt himself to be a teacher and expected to appear to others as such follows from a story he tells in a footnote in his *Analysis of Sensations*[21] (p. 4). Stepping into a bus in a tired state of mind he sees a man enter from the other side whom he identifies as an "old, shabby schoolmaster", when suddenly he recognizes that he is seeing himself in a big mirror. As he explained, "The physiognomy of my class, accordingly, was better known to me than my own."

VI

Essential ideas of Mach developed and were brought to fruition during his various teaching activities. One of his most important writings – and one which is least known (it is not included in George Sarton's *Horus*) – is a small book, the first literary fruit of his activity in Prague, which carries the rather longish title *History and Root of the Principle of the Conservation of Energy.*[19] Published in Prague in 1872 and again in Leipzig in 1909, it was translated into English in 1911. Mach stressed in his later writings that this little book contained many of his thoughts on the methods and aims of science; and indeed much of what he wrote later was in many ways a commentary on the thoughts expressed in this modest volume.

As the title indicates, Mach set himself the task of exploring the origin of the idea of conservation of physical energy. Both term and concept, were relatively new, as I mentioned before, but by 1872 they had become more or less accepted and Mach attempted now to show historically "that this theory is by no means as new as one tends to believe", and philosophically "that this theorem by no means stands and falls with the *mechanical* view of the world [and that] its logical root is deeper in our minds than that view". The book also contained the reprint of an earlier paper by Mach, dealing with the definition of mass, which he had unsuccessfully attempted to publish in the *Annalen der Physik* and which he had therefore placed in a less well-known journal, *Carl's Repertorium für physikalische Technik*, in 1868, and which he now included in this book on energy in order to bring its contents to the attention of the wider scientific community.

In this paper Mach had already proposed the idea that the laws of physics are descriptions and should not be conceived as causal relationships, anticipating Kirchhoff by two years. The most we can say about phenomena is that they depend on each other or that there exists a functional relationship between quantities we can determine. Laws of science, Mach defined, are limitations to our expectations under the guidance of our experience; thus the law has restrictive character, since any increase in scientific knowledge leads to a restriction of expectation. These expectations are the product of a psychological desire to orient ourselves in nature.

To the general discussion of the meaning of laws in science Mach formulates in this book another major thought, a corollary, namely the principle of scientific economy. With increasing accumulation of knowledge, scientists are motivated to obtain a reduction in intellectual effort and to achieve an economy in the structure of their laws. This principle has some resemblance to the *lex parsimoniae*, or 'Occam's razor', and to Newton's rules of scientific thinking; but for Mach the economy of thinking is conceived as an essential of our particular form of civilization. He had formed the idea in Graz under the influence of a colleague, Emanuel Herrmann, professor of economics, whom he quotes as having said that "science also has a problem of economy or thrift". In his book on heat (Ref. 11, p. 393) Mach discusses the economy principle also in connection with the Carnot coefficient of heat engines.

In one of the later editions of the *Mechanics* (Ref. 10, p. 591) Mach reminisces that the idea of economy of thought developed in him greatly through the experience of teaching and that, as a young lecturer, he flattered himself to have been the sole possessor of this truth. In the *Conservation of Energy* (Ref. 19, p. 88) he also said "that in the sciences ... we are chiefly concerned with the convenience and saving of thought I have maintained since the beginning of my career as a teacher". He again stressed the didactic connection when he wrote (*Mechanics*, p. 191): "The aim of instruction is simply saving experience; the labor of one man is made to take the place of that of another". But economy in teaching and in research is for him a method, not a doctrine of purpose. In the *Wärmelehre* (Ref. 11, p. 391), he wrote: "The methods through which knowledge is acquired are of an economic nature. What kind of use will be made of the acquired knowledge ... has nothing to do with scientific methods." Mach has been accused of regarding the economy of thinking as the *aim* of science, and on the basis of such misunderstanding a great physicist risked the gibe that Mach's economy principle would demand the suspension of all science, since "If we want to economize, the best way would be to stop thinking entirely".[20]

We are, in fact, approaching this ideal in our modern civilization by making use of computers, and with their help we may one day be able to abandon the formulation of laws; just store all our machine-recorded observations and when they are needed extract the specific information from the memory tapes. The economy of thrift of Mach's period, with its frugality in the home and a balanced budget in public life, has now been replaced by the Keynesian modern economy of planned credit, and we not only take scientific evidence on credit or faith, but are prepared to risk large expenditures of public finances on the outcome of scientific predictions, a procedure unheard of in Mach's time.

Mach's economy principle, properly understood, is a method to control the abundance or affluence of data which science produces and which seemed to him a desideratum for organizing the richness of scientific material. His was an economy in communication, a desire for that limitation of which Goethe said that it distinguishes the work of the true master. He did not puritanically demand that scientists should reduce the number of observables – the *entia* of Occam – and declare a moratorium on observations. On the contrary, as an investigator of optical and acoustical phenomena – he was also an accomplished pianist – Mach was a joyful observer of the world in all its manifestations, and the study of sensations was for him therefore not the least delightful of his manifold activities.

VII

Mach was, of course, aware that scientists did not only attempt to make and accumulate observations, but want to create unifying, objective theories. Such theories, he noted, unavoidably established a wall between the physico-physiological and psychological sciences if they took recourse to metaphysical concepts, as 'matter' and 'soul', or to speculative constructs, as 'atoms' and 'faculties'. Mach's interest in physics, physiology and psychology made it desirable for him to look for a common standpoint which would allow him to maintain it unchanged when turning from one field to the other. He believed he had discovered this common approach in the 'analysis of sensations', an idea which he developed in a book under this title[21], published first in 1885, two years after the *Mechanics*. The subtitle, *The relation of the physical and the psychical*, indicates that he tried to bridge the 'communication gap' between the physical and the psychological through a consistent phenomenological fixed point and the introduction of 'elements' of experience or sense perceptions. The interpretation of the term cannot be our task here, but it should be mentioned that its expansion in neo-positivistic philosophy led to the replacement of 'elements' by 'protocol sentences'.

Mach did not deny the value of independently existing special sciences. He was satisfied with having proposed an encyclopedic union which would shield the sciences from particular, superfluous concepts. The unity-of-science movements, the American religious monists, Wilhelm Ostwald's 'Energetics' and later the scientistic Vienna Circle, who claimed Mach as one of their own, far exceeded Mach's intentions; the Vienna Circle especially gave the impression of intolerance and militancy.[22] On the other hand, fanatical opposition to Mach, working both from scientific and political motivations, distorted his ideas by misrepresentation and insinuation. When Mach expressed himself against metaphysical assumptions, he did not mean to "explain away an objective world in terms of subjective ideas", as J. D. Bernal wants us to believe[23]; nor was he aware that thereby his philosophy played a role in the service of the bourgeoisie[24]. His rejection of metaphysics was not due to a lack of imagination or to the inability to invent 'constructs' or to appreciate models and symbols. In the arguments with Boltzmann on atomistics, he did not object to a discontinuous space concept as such, but only to the crude mechanistic picture of the atomists of 1899 (Ref. 11, p. 431).[25]

It should not be assumed that Mach's anti-metaphysical attitude originated from the analysis of sensations or the unity-of-science concept which served as its envelope. I believe that Mach's rejection of metaphysics was based on didactic and social experiences which convinced him that man's mind is incapable of reaching farther than simply to manipulate the elements of sensation. Or, to borrow a term from the 'analysis' of Mach's famous compatriot, it seems that he tacitly expected a 'Censor' to act in the human mind so as to restrict man's intellectual aspirations, as in Sigmund Freud's theory, a 'Censor' limits man's instinctive aspirations. In the field of scientific speculation Mach wished man to be passive, and in this way to achieve a perfect, man-oriented science for an imperfect mankind. Not unlike Michael Faraday (and Pierre Duhem) he kept the door between the science laboratory and the church firmly shut. Mach barred the door of the laboratory from within in order to protect science from the exuberance of human fantasy and emotionalism of any kind (Ref. 15, p. 88). He considered religious faith to be a private affair and he rejected any 'Concordat' between science and religion as well as the cosmic religiosity supported by Max Planck. His upbringing in a liberal and humane atmosphere saved him from the conflict of conscience which vexed many others, as for example Ernest Renan; Mach did not feel at any time the need for an *Examen de conscience*, as Renan did. On the other hand, in him the enlightened and pragmatic Austrian 'Josefinism' was tempered by the sereneness of the dominant spirit of the Austrian baroque.

Mach was never explicit about the motivations of his ideas and did not leave us with a *Summa* of his beliefs or a philosophical testament. Evidently there was no inclination in his disposition for scientific lyricism or pomposity, which easily go hand in hand with the creation of a synthetic philosophy and even of a positivistic system. By his own admission he liked to ramble through science; and he preferred to present his thoughts analytically in the form of short articles rather than through a deductive, comprehensive tome. With the exception of the *Mechanics* his books are collections of more or less independent essays on a variety of related topics. Occasionally he may have felt the lack of power of synthesis or of his own inability to subscribe to a definite doctrine; but, as he wrote in the *Mechanics* (Ref. 10, p. 559), he preferred "to endure an incomplete world view rather than to find satisfaction in a seemingly complete, but inadequate system".

VIII

Having recognized the wide range of Mach's interest it becomes obvious that Mach was neither a specialist's specialist, nor a generalist in the current sense. He originated several fields of research, but without exhausting their potentialities and he also extracted from his specialized studies general principles of new insight without claiming for them more than a relative validity for the 'human condition'. He attempted to spread the essence of his thoughts – which he refused to call a 'philosophy' –

through the spoken and the written word, but he never became dogmatic. His intuitive power of foresight led him to recognize future developments in the sciences. His inspirations were reactions rather than actions: reactions both to his reading in the history of physics and to classroom situations. He looked back to find the mistakes of the scientific past rather than to develop his insights into what he probably felt would be the mistakes of the future. As an historian of science he was not interested in teaching the development of science as part of a glorious, and as it now is often taught, a vainglorious 'Western tradition'. Basically, he was more interested in the questions that could be asked than in the answers which could be given now. The fundamental trait of his attitude was that of the enlightened teacher, of a man who shows others how to observe and to give meaning to their observations within self-imposed bounds; with the good teacher's understanding of the limitations of man's intellectual capacities.

Mach's papers and books, because of their discursive nature, are still capable of speaking to us with a new resonance of meaning. "Reading Mach's works", Albert Einstein wrote, "one feels the delight which the author had in the composition of his clear, concise sentences. However, not only the intellectual enjoyment and the appreciation of the easy style makes the reading of his books so attractive, but equally the kind, humane and hopeful convictions which shimmer between the lines when he is talking about what concerns mankind".[26] His position as a critic of scientific civilization is never marred by skepticism, though he sometimes shows a guileless irony. Just once did Mach lose his customary composure, and that was in his reply to Planck's unfortunate attack.[27] His only weakness is the tendency to repetition, probably in agreement with his didactic advice to teachers – according to a Prague tradition – in which he quoted from *Faust*, Part I: "Thrice must the words be spoken". He wrote for a public of scientists, avoiding in his style a verbosity which would more ingratiate than educate. He did not feign hypotheses and did not feign culture. With no obligation to the scientific 'establishment' or the public-at-large, he appears to us as the last liberal in the academic world who cherished freedom of thought much more than the safety and bondage of intellectual tenure.

I believe that much of the strength to go it alone came to Mach from his devotion to education. "Interest in education is only an expression of our interest in the world and in humanity", Herbart had written (Ref. 6,

p. 141). It was this constant human and social interest which gave motivation to the whole of Mach's activity, the understanding that he shared with his friend Popper-Lynkeus, of placing modest expectations on the intellectual as well as on the moral integrity of the individual man, but of expecting everything for human society from the establishment of workable and just institutions, science being one. Matthew Arnold observed about one hundred years ago that "while the leading humanists have brought their mind and energy to bear upon the teaching of their own studies, the leaders in the natural sciences, the Davys and Faradays, have not".[28] Arnold describes a situation which is still very much with us today when university scientists are expected to treat teaching quite casually. Mach was not the only exception to the rule, but he was an outstanding example of a scientist and teacher who gave equal weight to the two components of a science professor's daily tasks. I do not think that Mach's strong commitment to teaching and education could find many imitators today. Mach was too singular a personality and his activity too much bound to the conditions of his time to become an acceptable model in our time. The totality of his work, his experiments, his historical interests, his philosophical ideas and his humane personality will however long be remembered. And, as I hope to have shown, through a recognition of Mach's central concern for education we can best appreciate the full scope and depth of his thoughts and the origin of his many inspired reflections.

Vanderbilt University

LIST OF MACH'S EDUCATIONAL PUBLICATIONS

(Condensed from Joachim Thiele, 'Ernst Mach – Bibliographie', *Centaurus*, **8** (1963), 189–237.)

GENERAL PHYSICS TEXTBOOKS

Grundriss der Naturlehre für die unteren Classen der Mittelschulen. Ausgabe für Gymnasien, with Joh. Odstrčil and (beginning with third edition) K. Habart. Tempsky, Vienna und Prague, 1887, 231 pp. Seven editions, to 1910.

Ausgabe für Realschulen. Prepared by K. Habart. Tempsky, Vienna, 1896, 190 pp. Seven editions, to 1910.

Grundriss der Naturlehre für die oberen Classen der Mittelschulen. Ausgabe für Gymnasien. Tempsky, Prague, 1891, 320 pp. (Identical with Ref. 16 in References, below).

Ausgabe für Realschulen. Tempsky, Prague, 1892, 288 pp.

Ausgabe für Mädchenlyzeen (Girls schools). Prepared by K. Habart and Gallus Wenzel. Vienna, 1913, 130 pp.

Grundriss der Physik für die höheren Schulen des Deutschen Reiches. Prepared by Ferdinand Harbordt and Max Fischer. G. Freytag, Leipzig, 1893/4. I. Part. Preparatory course, 170 pp. II. Part. Advanced course, 180 pp. Four editions, the last in 1917.

ARTICLES

'Der relative Bildungswert der philologischen und der mathematisch-naturwissenschaftlichen Unterrichtsfächer der höheren Schulen', Lecture held at a teacher's conference, Dortmund, 1886. Tempsky, Prague, 1886, 22 pp. English translation: 'On the Relative value of the Classics and the Mathematico-physical Sciences in Colleges and High Schools', *Open Court* **8** (1894) 4295, 4308, 4311. Reprinted in *Popular Scientific Lectures* (Ref. 18, below).

The following published in *Zeitschrift für den physikalischen und chemischen Unterricht*, Springer Verlag, Berlin, **1–56** (1887–1943). 'Über den Unterricht in der Wärmelehre', **1** (1887/88), 3. 'Physikalische Denkaufgaben', **1** (1887/88), 211; **2** (1889/90), 28. 'Einfache Versuche über Strahlende Wärme', **7** (1893/94), 113. 'Über Gedankenexperimente', **10** (1896/97), 1.

REFERENCES

[1] Floyd Ratliff, *Mach Bands, Quantitative Studies on Neural Networks in the Retina.* Holden-Day Inc., San Francisco, 1965.

[2] K. D. Heller, *Ernst Mach, Wegbereiter der Modernen Physik.* Springer Verlag, Vienna, 1964.

[3] Anton Lampa, 'Ernst Mach', *Deutsche Arbeit* **15** (1915/16), 608.

[4] Felix Ehrenhaft, 'Fünfzig Jahre Physik im Spiegel der Wiener Chemisch-physikalischen Gesellschaft', Festrede. (Pamphlet), Vienna 1919; F. Herneck, 'Wiener Physik vor 100 Jahren', *Physikalische Blätter*, (1961), p. 455.

[5] Josef Popper-Lynkeus, *Mein Leben und Wirken, Eine Selbstdarstellung*, Carl Reissner, Dresden, 1924, p. 26.

[6] J. F. Herbart, *The Science of Education*, transl. by H. M. and Emmy Felkin, D. C. Heath & Co., Boston, 1896.

[7] Ernst Mach, *Compendium der Physik für Mediziner*, Wilhelm Braumüller, Vienna, 1863.

[8] The *Compendium* and several other textbooks by Mach were lent to me by the Ernst-Mach-Institut at Freiburg/Br. by kind permission of Dr. F. W. Merzkirch.

[9] Information on Mach's teaching activity in Prague has been collected for me by my old and dear friend Dr. Julius Kroczek to whom I express my best thanks. I also gratefully acknowledge the assistance given by Dr. Frant. Kavka, Director of the Archives of Charles University, Prague.

[10] Ernst Mach, *The Science of Mechanics. A Critical and Historical Account of Its Development.* Transl. by T. J. McCormack, New Introduction by Karl Menger, Open Court Publ. Co., La Salle, Ill., 1960.

[11] Ernst Mach, *Die Prinzipien der Wärmelehre. Historisch-kritisch dargestellt*, 4th edit., J. A. Barth, Leipzig, 1923.

[12] Ernst Mach, *The Principles of Physical Optics. An Historical and Philosophical Treatment.* Transl. by J. S. Anderson and A. F. A. Young, Dover Publ., New York, 1953.

[13] Cf. Mario Bunge. 'Mach's Critique of Newtonian Mechanics', *Amer. J. of Phys.* **34** (1966), 586.

[14] Josef Mayerhöfer, 'Ernst Mach as a Professor of the History of Science', *Proceedings 10th International Congress of the History of Science* (Ithaca, N.Y. 1962), Hermann, Paris, 1963, vol. II, p. 337.
[15] Ernst Mach, *Erkenntnis und Irrtum*, 5th ed., J. A. Barth, Leipzig, 1926.
[16] Ernst Mach, *Leitfaden der Physik für Studierende*, 2nd ed., Tempsky, Vienna, 1891.
[17] Mach was co-editor of the *Zeitschrift für den physikalischen und chemischen Unterricht*, from the first issue in 1887/88 to 1901, and wrote several articles for this journal. The editor was Dr. F. Poske in Berlin. Mach's successor in the co-editorship was the Austrian pedagogue Professor Alois Höfler. Höfler's Mach obituary was published in the *Zeitschrift* **29**, 57 (March, 1916) and stressed the importance of Mach's contributions in the field of science education.
[18] Ernst Mach, *Popular Scientific Lectures*, Transl. by T. J. McCormack, Open Court Publ. Co., Chicago, 1895, p. 285.
[19] Ernst Mach, *History and Root of the Principle of the Conservation of Energy*. Transl. and annot. by Philip E. B. Jourdain, Open Court Publ. Co., Chicago, 1911.
[20] Max Born, *Natural Philosophy of Cause and Chance*, Dover Publ., New York, 1948, p. 207.
[21] Ernst Mach, *Die Analyse der Empfindungen und das Verhältnis des Physischen zum Psychischen*, 7th edit. Gustav Fischer, Jena, 1918. (In English, Dover Publ. Co., New York, 1959.)
[22] Cf. on Mach and the Unity-of-Science movement: Philipp Frank, in *Journ. of Unified Science (Erkenntnis)* **7**: (1937/38) 247. [In this vol., pp. 235–244 – Ed.]
[2] J. D. Bernal, *Science in History*, Watts, London, 1954, p. 528.
[24] Article 'Mach', *Meyers Neues Lexicon*, VEB Bibliographisches Institut, vol. V, Leipzig, p. 527.
[25] L. Rosenfeld, 'Review of J. D. Bernal's *Science in History*', *Centaurus* **4** (1959), 285. Prof. Rosenfeld in this review quotes Stefan Meyer that Mach, on being shown, in 1903, the scintillations produced by alpha particles declared "Now I believe in the reality of atoms". This reads like one of the pious fabrications about a life-long disbeliever's return to the faith, *in extremis.* Professor Rosenfeld sees in Mach's remark a sign of his openmindedness. Since Mach did not mention his 'conversion' anywhere after 1903 it is hard to believe that he was serious when he spoke the words. Was he supposed to give up a founded scientific attitude in one moment of 'inspiration'?
[26] Albert Einstein, 'Ernst Mach', *Physikal. Zeit.* **17** (1916), 104.
[27] Max Planck, 'Die Einheit des Physikalischen Weltbildes', *Physikal. Zeit.* **10** (1909), 62. The controversy against Mach, p. 73ff; Ernst Mach, 'Die Leitgedanken meiner Naturwissenschaftlichen Erkenntnislehre und ihre Aufnahme durch die Zeitgenossen', *Physikal. Zeit.* **11** (1910), 599. Reply to Planck's Criticism, p. 603 ff. Max Planck, 'Zur Machschen Theorie der Physikalischen Erkenntnis', *Physikal. Zeit.* **11** (1910), 1186.
[28] Matthew Arnold, *Schools and Universities on the Continent*, Macmillan, London, 1868, p. 261.

FLOYD RATLIFF

ON MACH'S CONTRIBUTIONS TO THE ANALYSIS OF SENSATIONS*

The great diversity of Mach's interests and contributions is one of the most remarkable things about him. But most of us have seen only one, or – at most – only a very few of the many facets of his work. The cosmologist – of course – knows about Mach's Principle. A jet propulsion engineer is likely to know of Mach's pioneering work on supersonic phenomena and to utilize the concept of the Mach number in his daily work. Physiologists who study the history of their subject know about the Mach-Breuer-Brown theory of the function of the semicircular canals in the labyrinth of the internal ear. Psychologists working in the field of visual sensation and perception are familiar with the contrast phenomenon known as Mach bands. Mach's critique of Newtonian mechanics is well known to physicists. And the influence of this critique, and of Mach's positivistic views in general, on the development of relativity theory and on the course of modern science and philosophy is well known to philosophers and historians of science. But few psychologists have ever heard of Mach's Principle, and few cosmologists know that they see Mach bands every day. In short, most of us know of some significant contribution that Mach made in our own special field – or in one closely related to it – but it usually comes as a surprise to find that he made so many equally significant contributions in other fields.

In his old age Mach looked back over his life and career and – commenting on this diversity – spoke of himself as "merely an unprejudiced rambler, endowed with original ideas, in varied fields of knowledge". But his so-called rambles were neither aimless nor trivial; and rather than scattering his efforts, they unified them. Indeed, it was his alternate occupation with physics and the psychophysiology of the senses and his study of the historical development of the physical sciences that gave him such an unusual breadth of knowledge and enabled him to see, at first hand, threads of unity among scientific disciplines where others saw only diversity. Probably the one thing that set him apart from others of equal genius was this broader experience, and – based on it – his belief in the essential

Boston Studies in the Philosophy of Science, VI.

unity of science – indeed, the unity of all human knowledge. This belief rested mainly on the premise that all knowledge is based on experience and that no fact or truth may be established independently of experience. Mach was quite explicit about this. For example, in the preface of his work on *The Analysis of Sensations* he wrote:

> One and the same view underlies both my epistemologico-physical writings and my present attempt to deal with the physiology of the senses – the view, namely that all metaphysical elements are to be eliminated as superfluous and as destructive of the economy of science.

There is little doubt that Mach's antimetaphysical view of the theory of knowledge was largely conditioned by his very early interest – while still little more than a child – in the analysis of sensory experience. Furthermore, this view was undoubtedly reinforced from time to time throughout the whole of his long career by his continuing interest in, and numerous experiments on, sensory phenomena. For although his contributions to sensory physiology and psychology were almost as numerous and varied as were his contributions to other scientific disciplines, they nevertheless almost always centered on one aspect or another of this one fundamental problem: the role of the observer and his senses in the acquisition and shaping of our knowledge of the external world.

To illustrate this focus of Mach's interests, and to provide a background for a more general discussion later on, let us now consider four specific but diverse examples of his experimental and theoretical contributions to the analysis of the sensations. They are:

(1) on the alteration of the pitch of tones and the hues of colors by movement (the Doppler effect);

(2) on the role of the semi-circular canals in the internal ear in sensations of movement;

(3) on the interdependence of retinal points and its influence on the perception of brightness and form; and

(4) on the relation between physiological and geometrical space.

I. ON THE DOPPLER EFFECT

Several years before Mach entered the University of Vienna, Christian Johann Doppler had published his now famous theorem on the effect of the state of motion on observed frequency – the so-called Doppler effect.[1] According to the theorem both the observed hue of a color and the observed pitch of a tone should be affected by rapid movement of the source of the light or sound waves, or by movement of the observer himself, or by simultaneous movement of both source and observer relative to each other.[2] Specifically, as the distance between observer and source decreases, wavelengths should appear to shorten and frequencies to increase; as the distance between observer and source increases, wavelengths should appear to lengthen and frequencies to decrease.

Doppler's theorem was not immediately accepted by all of his contemporaries. Among those who did not accept it was Petzval, one of Mach's teachers at the university. Petzval's dissent was based on his own mathematical formulation of a 'principle of the conservation of period'. The essence of this principle was that a constant movement of the conducting medium relative to a fixed source and fixed observer should *not* affect the observed wavelength or frequency – that is, the perceived hue of a color or pitch of a tone should not be altered.

Doppler's theorem and Petzval's principle seemed to contradict each other – the one seeming to assert that the state of motion does have a significant effect on observed tones and colors, the other seeming to assert that it does not – and a controversy soon arose between the two men.

Doppler's theorem was originally based on the shift of spectral lines observed in the radiation of double stars (which alternately approach and recede from the earth as they rotate about their common center of gravity) and on changes in the pitch of locomotive whistles as speeding trains pass an observer. Petzval's principle was based on a purely mathematical analysis. Since neither Doppler's theorem nor Petzval's principle had grown out of experiments in which all significant variables could be carefully controlled, there was fertile ground for the controversy which arose. In one of his first researches to be published, Mach resolved this controversy. In a few simple and direct experiments he succeeded in demonstrating both the Doppler variation and the Petzval constancy in the laboratory and – at the same time – showed that there was no contradiction between the two. As it turned out, the whole controversy was based on a misunderstanding.

The apparatus Mach used in this work is illustrated in Figure 1. A six foot long bar *ab*, with an A tone whistle at the end *a*, was attached by a shaft *c* to the flywheel of a turning lathe (not shown) so that it could be rapidly rotated about the axis of the bearing *cd*. The right-hand side of the bearing was attached by an airtight packing box, at *d*, to a hollow tube *e*. A bellows was attached to the end of this tube at *f*. Thus, even during rotation, a continuous steady stream of air could be driven by the bellows into the tube *e*, through the hole in the axis of the bearing *cd*, and out

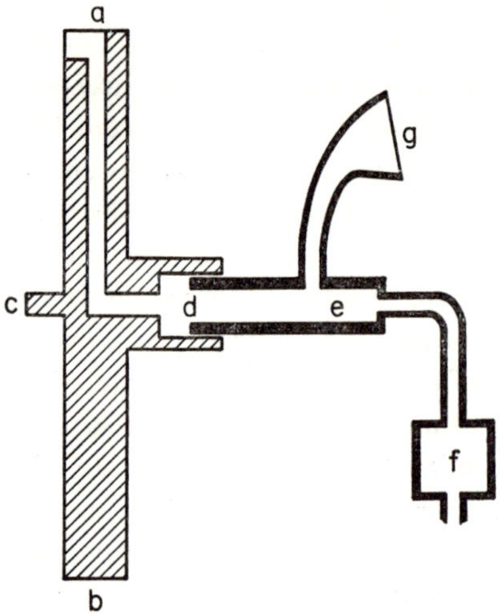

Fig. 1. Apparatus for experiments on the Doppler effect. (Redrawn after E. Mach)

along the hole in the bar *ab*, to the whistle at *a*. The observer could stand anywhere in the room and listen to the sound of the moving whistle conducted by the open still air. To enable the observer to hear the sound of the whistle conducted by the enclosed steady current of air, a small tube was attached to the side of the main tube *e*, and its flared outer end covered with an elastic membrane *g*.

When the observer stood in the plane of rotation of the bar *ab*, and listened to the sound in the open still air, he heard the pitch of the tone rise and fall as the whistle moved toward and away from him – as it should have, in accordance with Doppler's theorem. If the speed of rotation was increased, the difference between the highest and lowest pitch of the tone became greater; if decreased, it became smaller. If, however, he stood somewhere along the axis *cd*, so that there was no change in the distance from his ear to the rotating whistle, there was no change in the pitch of the tone.

If the observer now put one ear to the diaphragm and stopped up the other, the pitch of the sound of the whistle borne to his ear over the moving current of air was constant, since the distance to the whistle was constant even during rotation. But not only was the pitch of the tone conducted to the ear over the moving air constant, it was also essentially the same as that conducted directly over a fixed distance from the whistle to the ear through the open still air. That is, the pitch of the sound borne to the ear in the one case by still air and in the other by a steady stream of air of considerable speed in the direction of the source was the same – an illustration of Petzval's principle. Furthermore, by placing one ear to the diaphragm and leaving the other ear open, the observer could hear, at one and the same time, the constant pitch of the tone conducted through the tube and the rise and fall of the pitch of the tone conducted through the open air.

Thus Mach showed that Doppler's theorem and Petzval's principle were complementary rather than contradictory; Doppler was concerned with motion of the source and observer relative to one another – Petzval with motion of the conducting medium relative to a stationary source and a stationary observer. Both were essentially correct; they merely treated different aspects of the same problem.

To the mathematician and the physicist, all this might seem to have been nothing more than an experiment on the physics of wave motion. But it was in fact a *psychophysical* experiment. The variations in the pitch of the tone were determined by purely subjective means; Mach had no physical instruments for measuring the frequency of the sound waves. The sense of hearing was absolutely essential, and the vantage point of the observer was critical. No doubt it was through such experiments – carried out very early in his career – that Mach gained his penetrating insight into the indispensable central role of the observer and his senses in the investigation of physical phenomena.

II. ON THE ROLE OF THE SEMICIRCULAR CANALS OF THE INTERNAL EAR IN SENSATIONS OF MOVEMENT

A chance observation during a trip on a train called Mach's attention to the problem of the function of the semicircular canals of the internal ear. He noticed that trees, houses, chimneys, and other fixed objects seemed to swerve from the vertical when he rounded a sharp curve. The generally accepted explanation was extremely simple: we regard the railway car as remaining vertical and unconsciously infer the inclination of the trees. Unfortunately, as Mach pointed out, the opposite conclusion that we regard the trees as vertical and infer the inclination of the car, is equally clear on this theory. Mach reasoned that this and many other related phenomena might easily be explained on the assumption that an organ exists in the head which responds to accelerations, and by means of which we are made aware of movements. (Breuer and Kreidl were to prove later that the optical distortion of the vertical – likely to be experienced by anyone seated in a whirling chair, or in a railway coach passing rapidly around a sharp curve – depends upon an involuntary reflex wheel-like movement of the eyes.)

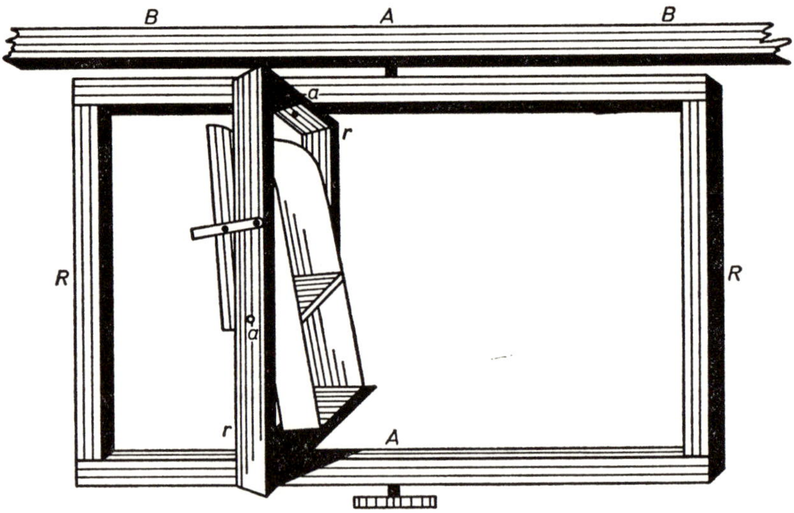

Fig. 2. Rotating chair for the study of sensations of movement. (E. Mach[2])

Mach immediately set to work to investigate the characteristic sensations which accompany active and passive motions of the body. To do this he constructed a huge rotating chair, shown in Figure 2, in which an observer seated upright could be rotated about his own vertical axis *ae*, or about an axis *AA* eccentric to his body, or any combination of the two. The subject could be enclosed in a lightweight paper box (not shown in the figure) in order to eliminate external cues. This chair was a gigantic instrument constructed out of 4 × 4 inch timbers over four meters long, but it would be dwarfed by the enormous centrifuges (for which it was the prototype) that are now used for training astronauts.

In his experiments on the passively moved human Mach found among other things that – without external cues – we have no direct perception of speed of rotation, only of acceleration and deceleration. Thus, with external cues removed, rotation is perceived during the acceleration of starting rotation from rest, there is no perception of rotation when speed of rotation is steadily maintained, and deceleration or stopping yields the perception of reversed rotation. (Analogous effects are observed, of course, during linear acceleration and deceleration.)

Since we perceive acceleration and deceleration rather than 'absolute' motion, states of rest and of uniform motion are often confounded – even when there are apparently adequate visual cues to distinguish between the two. For example, it is a common experience for passengers on the slowly and uniformly moving Staten Island Ferry approaching the pier at the Battery of New York City to have the impression, when the harbor is very calm, that the ferry is stationary and that the pier and lower Manhattan skyline are approaching it. Everyone, probably, has had the converse experience of a sensation of movement while standing still on a bridge and looking down at the flowing water.

Mach proposed that inertia of the fluid contents of the semicircular canals in the labyrinth of the internal ear could adequately account for both his experimental findings and everyday experiences of the kind just described. With the starting of rotation of the body from rest, for example, the fluid would tend to remain at rest and would thus press against receptors in the ampullae and thereby excite the neurons connecting these receptors with the brain. During uniform rotation the fluid would gradually come to rest with respect to the canals just as it does when the body is stationary for some time. Upon deceleration or stopping rotation, the free fluid would continue to move, thus generating the same sensation as starting of rotation, or as an acceleration, in the opposite direction. Of course, our sensations of movement are not restricted to the horizontal plane. The two sets of semicircular canals, being symmetrically arranged with each canal in each set in a different one of three intersecting planes, can provide – in combination – precise information about rotation in any direction.

Mach confirmed his work on human subjects by observing the behavior of a number of small vertebrates (birds and fishes) in a specially constructed instrument, in similar situations. Independently, Breuer (in Vienna) and Brown (in Edinburgh) reached conclusions similar to Mach's and the general theory outlined above is often referred to as the Mach-Breuer-Brown theory. While not correct in all details, as originally formulated, numerous anatomical, electrophysiological, and psychophysical experiments have since shown that the Mach-Breuer-Brown theory is correct in its essential features.

Nowhere is Mach's application of physical concepts to the study of sensory processes better illustrated than in his work on the function of the semicircular canals. As a matter of fact, it was his knowledge of mechanics and of fluid dynamics which gave him clear insight into the nature of the problem and led to its quick solution. But the paths from one scientific discipline to another are not one way. Indeed, it is reasonable to believe that ideas arising from these experiments by Mach on the sensations of movement were not far removed from, and may well have led to, some of the ideas underlying his critique of the physical concepts of absolute space and motion.

III. ON THE INTERDEPENDENCE OF RETINAL POINTS AND ITS INFLUENCE ON THE PERCEPTION OF BRIGHTNESS

Upon completion of his doctoral studies at Vienna Mach intended to experiment in the field of physics, rather than in psychophysiology. But he lacked sufficient money to buy the relatively expensive equipment needed for research in physics. For this reason – more than for any purely scientific considerations – he turned for a time to the study of vision. His own eyes provided the necessary optical equipment, and for stimulus control he could use simple rotating discs.

As is well known, a white disc with a black sector appears gray when rapidly rotated – light gray if the black sector is small, dark gray if the black sector is large. Thus, by painting the appropriate black and white sectors on a disc, and then rotating it at high speed, one can obtain any desired distribution of illumination from the center to the circumference of the disc.

The apparent brightness of a succession of light and dark periods, at frequencies above the flicker-fusion-frequency, is given by the so-called Talbot-Plateau law. In brief, the law states that the apparent steady brightness of a very rapidly rotating disc with black and white sectors is equal to the mean value of the real varying brightness. But Mach noticed, by chance, that certain spatial distributions of illumination did not seem to obey the Talbot-Plateau law. Two of these, and the discs used to obtain them, are shown in Figure 3.

The calculated luminance along any radius of the upper disc, when set in rapid rotation, is uniformly high from the center of the disc (α) out to the point (β). It then decreases uniformly to a point (γ). Beyond this point it decreases still more rapidly out to the circumference. According to the Talbot-Plateau law, then, the rotating disc should have appeared a uniform bright gray from α to β, from β to γ it should have appeared uniformly darker and darker, and from γ to the circumference it should have appeared increasingly still darker.

The disc did appear gray during rotation, and – in general – each of the three zones appeared more or less as expected, as shown by the shading of the disc at the right. The transition from each zone to the next, however, was interrupted by a distinct narrow ring – a bright ring at the point β and a bright ring at the point γ.

The lower disc is the negative of the upper one and accordingly appeared, during rotation, as shown at the right. Where the 'positive' disc appeared bright, the 'negative' disc appeared dark. Accordingly, dark rings – instead of bright rings – were seen at β and γ. These striking and completely unexpected bright and dark rings are now known as Mach's rings or, more generally, *Mach bands*.

Mach investigated many different distributions of illumination and found that, in general, wherever there is a concave flexion in the light curve that part of the distribution appears brighter, wherever there is a convex flexion in the light curve that part appears darker. Indeed, once one's attention has been called to the Mach bands, one can see them everywhere. For example, shadows cast by almost any object in sunlight yield a distribution of illumination with both a convex and a concave flexion and exhibit the corresponding Mach bands. The full shadow (umbra) provides the uniform dark field, the half-shadow (penumbra) provides the graded field, and the area in full sunlight provides the uniform bright field. A dark band appears at the boundary between the full shadow and the half-shadow, and a bright band appears at the boundary between the half-shadow and the full sunlight. There are no such maxima and minima, of course, in the actual distribution of the light measured by a photometer or similar physical instrument. The Mach bands can seemingly be photographed, but this does not mean that they are necessarily physical phenomena. The camera is simply reproducing the original distribution of light. The photograph, like the original distribution, must be viewed by eye for the Mach bands to appear.

The Mach bands are one of the most compelling of all the visual 'illusions', and have been mistaken for objective physical phenomena on a number of occasions. One notable example of such a mistake occurred in an early investigation of the properties of X-rays; the subjective Mach bands were mistaken for the maxima and minima of a diffraction pattern.[5]

Mach deduced that the bright and dark bands resulted from an interdependence of the light sensitive elements (rods and/or cones) in the neural network formed by the retina. He assumed that each element transmitted information via the optic nerve to the brain about the intensity of illumination on it by means of changes in its level of excitation. (We now know that this is done by a frequency code, that is, by means of the number of nerve impulses discharged in the optic nerve per second. In general, the brighter the light on a particular receptor, the greater the frequency of impulses discharged by it.) Mach also assumed that each element exerted an inhibitory influence on its neighboring elements – an influence which reduced their levels of activity, and thus lessened the excitation that they sent to the brain. The brighter the light, the greater the inhibition. Furthermore, Mach assumed that the inhibition depended on distance: strong inhibitory effects being exerted on near neighbors, weak effects, or no effects at all, on distant ones. Since each element was a neighbor of its neighbors, they inhibited one another – that is, the inhibitory effect was mutual.

The physiological and psychological significance of the dependence of the inhibition on distance may be readily understood. Since intensely stimulated receptors exert a stronger inhibition on less intensely stimulated ones than the latter exert on the former, and since these effects diminish

Fig. 3. Discs for investigations on the interdependence of retinal points. (E. Mach[5])

with distance, contrast will be most strongly enhanced near the borders of differently stimulated regions of the receptor mosaic. As a result we see bright and dark bands in those regions of light patterns where the second derivative of the physical light curve has its minima and maxima.

Mach's insight into the nature and significance of the opposed processes of excitation and inhibition in the nervous system, along with his attempt to treat them in precise mathematical terms, was a major contribution to physiological psychology. It is most remarkable that Mach was able to deduce some of the properties of these fundamental processes nearly three quarters of a century before we were able to directly observe these same properties by electrophysiological techniques. His work was so far ahead of the times, however, that its significance was not recognized until recently. Indeed, it was not until 1928 that attention was called to Mach's contribution in this area by Prof. Georg von Békésy, who saw the possible relevance of it to his studies on the neural mechanisms for the sharpening of the perception of pitch in hearing.

The analysis of the role of these basic processes in the functioning of all parts of the nervous system now occupies some of the most prominent neurophysiologists of our day, and we need not dwell on the significance of these fundamental processes of excitation and inhibition for modern neurophysiology. The interplay of excitation and inhibition in our own brains is, no doubt, the physiological basis of all experience and thought. Indeed, in one of his papers[5] on the interdependence of retinal points Mach remarked that:

> We cannot understand the state before we understand the man. Likewise, we will not understand the psychology of man before we understand the much simpler psychology of the ganglion cells. The similarity of this psychology with that of the whole man permits conjecture and may serve us as a guiding star. It may thus well be the main task of physiological psychology to solve these simple basic psychological processes in the ganglion cells, out of whose reciprocal actions arises the psychology of man.

IV. ON THE RELATION BETWEEN PHYSIOLOGICAL AND GEOMETRICAL SPACE

The space of the Euclidean geometry that we have developed over many hundreds of years of deliberate experience and training is everywhere and in all directions constituted alike; it is unbounded and it is infinite in extent. But the visual space which we all come to know in the natural course of events in the short span of our individual lives is neither infinite in extent, nor unbounded. Thus our immediate perception of space, which Mach called 'physiological space', is considerably different from geometrical space.

Indeed, two very simple square figures may be geometrically congruent, but physiologically quite different, as is illustrated by the two nearby squares in Figure 4. These are seldom, if ever, immediately recognized as

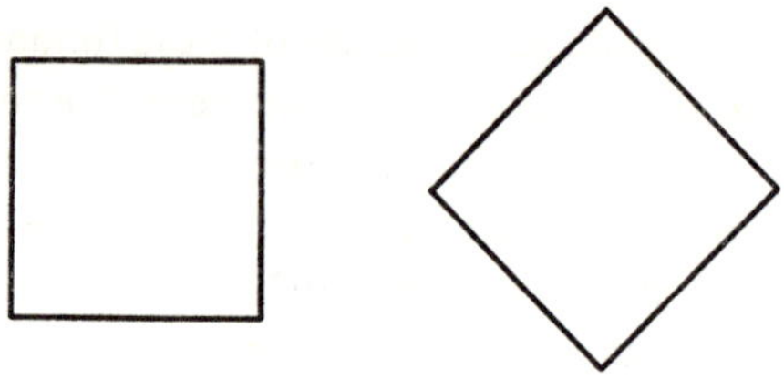

Fig. 4. Two figures with geometrically congruent forms, but different visual appearances. (E. Mach[4])

being geometrically the same. Usually their identity is perceived only after some intellectual or mechanical operations.

Obviously, says Mach, the *orientation* of a form, relative to the orientation of our own bodies is most significant physiologically. Our biological needs would not be satisfied with the pure relations of geometrical space. Three cardinal directions are plainly marked on us: above and below our bodies are unlike, as they are also in front and behind and – to a lesser extent – to the right and to the left. Sensations that make us aware of the differences between above and below, front and back, right and left (and far and near, also) must therefore have different qualities. It is the locality of an object with respect to our own bodies, not merely its locality with respect to other objects, that is biologically significant.

The ease with which we recognize the identity of simple forms in various locations and orientations and the extent to which we distinguish or confuse symmetrical forms seems definitely related to the corresponding orientation and symmetry of our own bodies. For example, the similarity of two forms having vertical symmetry is easy to recognize, and the similarity of two forms having horizontal symmetry difficult to recognize because – according to Mach – vertical symmetry produces a repetition of nearly the same sensation while horizontal symmetry does not. The fact that forms symmetrical about the vertical (median) plane are so often confounded, he maintained, is evidence that they produce the same or very similar sensations. Otherwise one would recognize them as different.

Mach cites the frequent confounding of certain letters of the alphabet in support of this view. Although made up of apparently simple plane figures, the various forms of the letters of the alphabet are rather complex and have several different and interesting kinds of symmetries.

A simple bit of parlor magic which mystifies children (and adults – sometimes) illustrates some of the peculiar symmetries of the capital letters. A horizontal mirror image of the slogan CHOICE QUALITY, which appears on the side of packs of certain cigarettes, is formed by laying the pack flat on the surface of a mirror. The word CHOICE is perfectly legible, but at the same time – in the same image – the word QUALITY is illegible. This results, of course, from the horizontal symmetries of some of the capital letters.

Of the twenty-six large characters of the alphabet, eleven are vertically symmetrical (A, H, I, M, O, T, V, W, X, Y). That is, each appears the same when turned about its vertical axis. Four of these eleven and five others are horizontally symmetrical, or nearly so (H, I, O, X, and B, C, D, E, K). Three are centrally symmetrical (N, S, Z), and the remainder are unsymmetrical (F, G, J, L, P, Q, R).

Mach points out that children learning to read and write the capital letters, and many semi-literate adults, often make errors having to do with handedness, i.e., errors about the median plane. Mirror writing is frequently seen in the writing of young children – E written as Ǝ, B as ᗺ, etc. N, S, and Z are frequently turned right to left; one often sees these latter errors in home-made roadside signs.

The lower case letters, of course, are quite different. Let us consider the following example of right-left and up-down symmetry:

d b

q p

Children just learning to read and write very frequently confound **d** and **b**, and **q** and **p** – but almost never **d** and **q**, or **b** and **p**. (Neither do they confound **d** and **p**, and **q** and **b**.) Notice that **d** and **b** and **q** and **p** are the two halves of a vertically symmetrical figure, while **d** and **q** and **b** and **p** are two halves of a horizontally symmetrical figure. On the assumption that confusion is only possible of things in us that excite the same or similar sensations, says Mach, it is evident that the sensations associated with right and left are much less different than are those associated with up and down and front and back. (As to right and left handedness itself – any drill sergeant can testify to the difficulty in teaching a recruit to know his right hand from his left, and – especially – his right foot from his left.)

It is a fact that the two halves of a vertically symmetric figure (including the two halves of our own bodies) are easily confounded and the assumption that they therefore probably produce very nearly the same sensations is reasonable. The question then arises, *why* should the two halves of a vertically symmetrical figure produce the same or similar sensations? According to Mach the answer is: because our apparatus of vision is itself vertically symmetrical.

Whatever external resemblances one eye (and the part of the nervous system associated with it) may have with the other, they are not identical. The right eye cannot take the place of the left one. This can easily be demonstrated by means of Wheatstone's pseudoscope – a simple optical device which interchanges the roles that our eyes play – the left eye seeing from the point of view of the right eye, and the right from the point of view of the left. Now we enter a new world: what is convex appears concave; what is concave, convex. What is far appears near and what is near appears far. The two eyes cannot be interchanged by such a mere lateral displacement because one eye is the *vertical* reflection of the other, both in structure and in function. Indeed, each half of the entire visual system – all the projections from each retina to the visual cortex – are but vertical mirror images of one another. Is it any wonder, then, asks Mach, that forms in space that are vertical reflections of one another – that is, the two halves of a vertically symmetrical figure – should be confounded?

Mach's explanation seems reasonable. Symmetries and ambiguities that he saw in the structure and function of the visual system as a whole can now be observed in the structure and function of its neural components. Although many retinal ganglion cells and many neurons of the visual cortex are highly specialized to respond to particular spatial and temporal configurations of the pattern of light and shade on the retina, their responses may nonetheless be ambiguous. The symmetry of their structure is often such that mirror images of the particular spatial and temporal configurations which excite them may yield identical responses. Evidently, as Mach suggested, symmetry of structure and ambiguity of function go hand in hand.

V. DISCUSSION

Perhaps one quarter – or more – of all of Mach's published work dealt with some aspect of the analysis of sensations, and there was hardly a major area in the field to which he did not make a significant contribution at one time or another during his long and productive career. It would be presumptuous, therefore, to attempt either to survey or to evaluate the whole of his work on the senses in one short paper. The few specific examples, however, of his experiments on, and theoretical treatments of, the senses that we have just considered do adequately characterize some of the most significant and distinctive features of his approach to the analysis of sensations. In my view these are:

(1) the rigorous application of physical concepts and techniques and mathematical modes of thought to the study of physiological psychology;

(2) an organic or holistic view of the functions of the senses;

(3) the application of the idea of evolution to the theory of the sense organs; and

(4) the central role played by sensory experience in his theory of knowledge.

Let us now consider briefly, each of the above four features of Mach's work on the senses. In order to present Mach's own views – rather than an interpretation of them – a number of his own statements on these matters will be quoted.

A. *Biophysics*

Mach's training and experience as a physicist led him, quite naturally, to view biological problems in the framework of physics and mathematics. This view encompassed not only the obvious physical aspects of the bodies and behaviors of organisms (such as the inertia of the fluids in the semicircular canals), it included also those aspects which were then regarded as purely psychical. For example, a distinguished scholar once asked Mach about his work on visual contrast effects: "Why bother with this? It must be only an illusion." Mach's reply was that even 'illusions' are facts, and that laws underly them. And in his attempts to determine these laws, he did not hesitate to apply physical concepts and mathematical techniques wherever it seemed appropriate to do so. But – unlike some of our contemporaries – Mach did not envisage an easy and immediate solution to all biological problems by the mere application of physical techniques and mathematical methods of analysis. On this matter he wrote:

Every organism together with its parts is subject to the laws of physics. Hence the legitimate attempt gradually to conceive of an organism as something physical, and to establish the consideration of it in a 'causal' point of view as alone valid. But whenever we try to do this we are always brought face to face with the peculiar characteristics of the organic, for which no analogy can be found in the physical phenomena of 'lifeless' nature, so far as they have been investigated at present.

We do not indeed know what are the physical counterparts to memory and association. All the explanations that have been attempted are very much forced. In this respect it seems as if there were almost no analogy between the organic and the inorganic. It is possible, however, that in the physiology of the senses, psychological observation on the one side and physical observation on the other, may make such progress that they will ultimately come into contact, and that in this way new facts may be brought to light. The result of this investigation will not be a dualism but rather a science which, embracing both the organic and the inorganic, shall interpret the facts that are common to the two departments[4] (pp. 100–101).

B. *Holism*

Mach's organic, or holistic, view of the functional organization of the sensory systems, particularly that of the visual system, anticipated many of the points of view advanced by the Gestalt school of psychology. Concepts of 'fields' of influence exerted among interacting elements in the nervous system, the numerous analogies with complex physical systems, and the primacy of the whole over the part, all give Mach's work a 'Gestalt' flavor. Yet in no case does he claim that the whole is in any mysterious way 'greater than the sum of its parts'. His view was simply that the parts of sense organs, acting in concert, were able to carry out certain functions that they could not carry out individually. In his fourth article on the retina he wrote:

What appears to me to arise mainly from these phenomena and observations, above all, is that perception and even sensation are the result of countless individual organic forces and that we must regard the retina as more vital and more independent than we have been accustomed to doing. In recent times the physical view has passed through a not dissimilar change. In place of the concept of a gas there appeared the concept of an animated confusion of molecules whose uniform actions only arise from the laws of probability. The concept of a liquid appears to be changing similarly, and soon one will find that crystals cannot be regarded as being overly simple and uniform.

Since every retinal point perceives itself, so to speak, as above or below the average of its neighbors, there results a characteristic type of perception. Whatever is near the mean of the surrounding becomes effaced, whatever is above or below is disproportionately brought into prominence. One could say that the retina schematizes and caricatures. The teleological significance of this process is clear in itself. It is an analog of abstraction and of the formation of concepts.

Mechanics, as is known, always leads especially to problems of minima and maxima. The significance of these minimum principles, however, goes far beyond the special hypotheses of mechanics. They come into play wherever several factors act together, partly facilitating one another, partly inhibiting one another[5] (p. 299).

C. *Evolution*

Throughout his writings on the senses Mach made frequent references to the adaptive or evolutionary significance of various sensory processes. Examples of this may be found even in his earliest papers published in the 1860's. Mach's ready acceptance of the basic ideas of Darwin's theory (which had been published only a few years earlier, in 1859) is not surprising – his father had been much interested in animal behavior and Mach was thoroughly familiar with the pre-Darwinian ideas on evolution. Furthermore, evolutionary ideas were not foreign to Mach's own thinking. For example, he regarded the competition of conflicting concepts in science as having an adaptive value – being one of the means by which science gradually evolves and adjusts itself to changing times and to new ideas and information. Concerning teleological ideas in general he wrote:

Teleological conceptions, as aids to investigation, are not to be shunned. It is true, our comprehension of the facts of reality is not enhanced by referring them to an unknown World-Purpose, itself problematical, or to the equally problematical purpose of a living being. Nevertheless, the question as to the value that a given function has for the existence of an organism, or as to what are its actual contributions to the preservation of the organism, may be of great assistance in the comprehension of this function itself. Of course we must not suppose, on this account, as many Darwinians have done, that we have 'mechanically explained' a function, when we discover that it is necessary for the survival of the species. Darwin himself is doubtless quite free from this short-sighted conception. By what physical means a function is developed, still remains a physical problem; while the how and why of an organism's voluntary adaptation continues to be a psychological problem. The preservation of the species is only one, though an actual and very valuable, point of departure for inquiry, but it is by no means the last and the highest. Species have certainly been destroyed, and new ones have as certainly arisen. The pleasure-seeking and pain-avoiding will, therefore, is directed perforce beyond the preservation of the species. It preserves the species when it is advantageous to do so, and destroys it when its survival is no longer advantageous[4] (pp. 80–82).

In his work on the relation between physiological and geometric space Mach refers again and again to the biological adaptation of sensory processes and their evolutionary significance. For example:

The biological needs would not be satisfied with the pure relations of geometric space. 'Rightness', 'leftness', 'aboveness', 'belowness', 'nearness', and 'farness', must be distinguished by a sensational quality. The locality of an object, and not merely its relation to other localities, must be known, if an animal is to profit by such knowledge. It is also advantageous that the sensational indices of visual objects which are near by and consequently more important biologically, are sharply graduated; whereas, economy is practiced with the limited stock of indices at hand in the case of remote and less important objects'[3] (pp. 11–12).

If sensation generally, inclusive of sensation of space, be conceived not as an isolated phenomenon, but in its biological function, in its biological relationship, the entire subject will be rendered more intelligible[3] (p. 16).

As we recognize no real gulf between the physical and the psychical, it is a matter of course that, in the study of the sense organs, general physical as well as special biological observations may be employed. Much that appears to us difficult of comprehension when we draw a parallel between a sense organ and a physical apparatus, is rendered quite obvious in the light of the theory of evolution, simply by assuming that we are concerned with a living organism with particular memories, particular habits and manners, which owe their origin to a long and eventful race-history[4] (p. 71).

D. *The Foundations of Science*

Many scientists – perhaps most – give little or no thought to the foundations of their work. And when they do give some thought to the matter they are likely to disagree with Mach, either explicitly or implicitly. Consider, for example, the following statement attributed to a well-known physicist, Professor Wolfgang K. H. Panofsky:

All other physical, and probably all life sciences must ultimately rest on the findings of elementary particle physics. It would indeed violate all our past experience in the progress of science if nature had created a family of phenomena which governs the behavior of elementary particles without at the same time establishing any links between these phenomena and the large-scale world which is built from those very particles.[6]

For Mach, however, physical bodies, substances, atoms (and 'particles', too – had they existed in his time) were but thought symbols for complexes of sensations. Such symbols, concepts, and theories, he believed to be only provisional, and subject to change or even abandonment whenever the facts so warranted. Indeed, he once wrote that they "are like dry leaves which fall away" once they have served their function. Reality for Mach, if there was any to be found, was in the complex of sensations on which theoretical concepts were based – not in the concepts themselves. Concerning his contributions to the analysis of sensations he wrote:

The frequent excursions which I have made into this province have all sprung from the profound conviction that the foundations of science as a whole, and of physics in particular, await their next great elucidations from the analysis of the sensations.[4] (pp. vii–viii).

and on physics and psychology (p. 45):

The physicist says: I find everywhere bodies and the motions of bodies only, no sensations; sensations, therefore, must be something entirely different from the physical objects I deal with. The psychologist accepts the second portion of this declaration. For him, as is proper, sensations are the primary data; but to these there corresponds a mysterious physical something which, conformably with the prepossession, must be quite different from sensations. But what is it that is the really mysterious thing? Is it the Physis or the Psyche? Or is it perhaps both? It would almost appear so, as it is now the one and now the other that appears unattainable and involved in impenetrable obscurity. Or are we here being led around in a circle by some evil spirit?

I believe that the latter is the case. For me the elements [i.e., the sensations] ... are immediately and indubitably given, and for me they can never afterwards be volatilized away by considerations which ultimately are always based on their existence.

Rockefeller University

REFERENCES

* The preparation of this paper was supported, in part, by a grant (NB 00864) from the National Institute of Neurological Diseases and Blindness, U.S. Public Health Service. Acknowledgments are due to Mrs. Maria Lipski who translated some articles from the German.

[1] Mach, E., *Beiträge zur Doppler'schen Theorie der Ton- und Farbenänderungen durch Bewegung*, J. G. Calv'sche k. und k. Univ. Buchhandl. (Ottomar Beyer), Prag, 1873.

[2] Mach, E., *Grundlinien der Lehre von den Bewegungsempfindungen*, Wilhelm Engelmann, Leipzig, 1875.

[3] Mach, E., *Space and Geometry* (transl. by T. J. McCormack), The Open Court Publishing Company, Chicago, 1906.

[4] Mach, E., *The Analysis of Sensations and the Relation of the Physical to the Psychical* (transl. by C. M. Williams, revised by Sidney Waterlow), The Open Court Publishing Company (Chicago and London, 1914).

[5] Ratliff, F., *Mach Bands: Quantitative Studies on Neural Networks in the Retina*, Holden-Day Inc., San Francisco, 1965.

[6] Taylor, H., "From 'the Lynxes' to Stanford's 'LIN-AC'", *American Scientist* **54** (1966), 333–344.

WOLFGANG F. MERZKIRCH

MACH'S CONTRIBUTION TO THE DEVELOPMENT OF GAS DYNAMICS*

In these days the name of Ernst Mach is most often related to the well-known Mach number, a quantity which has become very popular since flying at supersonic speeds has developed so rapidly. If one takes a look at a textbook of aerodynamics or gas dynamics the name Mach appears in connection with other terms: Mach angle, Mach wave, Mach reflection are frequently used aerodynamic expressions. Ernst Mach, indeed, has provided the fundamental research for explaining and defining these terms, and is considered, therefore, with good reason, to be one of the originators of supersonic aerodynamics. In his ballistic experiments he discovered the existence of a bow-shock in front of a flying projectile and, thereby, an essential difference between flights at subsonic and supersonic velocities; he was the first to recognize the real nature of shockwaves in air and their 'irregular' reflection pattern; he visualized particular wave phenomena in a supersonic free-gas jet; and he originated the experimental techniques which were necessary to perform those investigations.

Most books and papers, however, written about the person of Ernst Mach, barely give reference to this gas-dynamic research, though it led to 18 original publications by Mach (Refs. 1 through 18). What has mostly been reviewed, analyzed and criticized is the importance of Mach's contributions to the fundamentals of science, the physical and philosophical value of his work, and the discussion of these questions has not yet been brought to an end. Mach's gas-dynamic research is, compared to the foregoing, a chapter that has found a certain completion, due to the particular scope of this field. It is true that Mach did not bring these problems to a definitive scientific solution, but his findings are correct in themselves because they are statements of facts which can be objectively perceived.

At this point, however, the relation between Mach's gas-dynamic research and his total work becomes evident: the results are facts perceptible to the human eye, since Mach succeeded in making visible aerodynamic phenomena which had been invisible or even unknown up to that time; he made it possible for them to be recognized physiologically, and thereafter found a correct physical interpretation. This was definitely a success for Mach's methodology that failed, however, when applied to other fields in physics, and the critics of Mach usually refer to those failures. A physical theory, of course, is disproved by only one sample-case of opposite evidence; but by studying Mach's success in the aerodynamic field one might find an easier access to appreciate Mach's way of thinking. Mach was primarily an experimentalist; he wanted to perceive with his senses the implications of physical statements, laws, and rules.

> Modern science strives to construct its picture of the world not from speculations but so far as possible from facts. It verifies its constructs by recourse to observation. Every newly observed fact completes its world-picture, and every divergence of a construct from observation points to some imperfection, to some lacuna in it. What is seen is put to the test of, and supplemented by, what is thought, which is again naught but the result of things previously seen. It is always peculiarly fascinating, therefore, to subject to direct verification by observation, that is, to render palpable to the sense, something which we have only theoretically excogitated or theoretically surmised.[18]

The aerodynamic phenomena associated with flying projectiles and the unsteady shockwaves are not visible to the naked eye. Mach made them visible by means of an excellent experimental technique, and he found the correct interpretation of the visualized features which he admitted then to be facts. He did not take it as his mission to give the theoretical analysis of the visualized phenomena; but his experimental findings became the fundamental background for many theoretical investigations by subsequent scientists.

Mach's studies in the field of gas dynamics, regarded in this way, are no isolated chapter within his total work. They are, of course, of minor importance compared to his contributions to the fundamentals of science, to the development of positivism and phenomenology. But they might enable us to gain an appreciation of the over-emphasized sensualism that Mach later turned to and of his unsuccessful experimental efforts of his last years in the field of optics.

Unfortunately little is known of what Mach himself thought of his gas dynamic research. He had collected a considerable amount of literature and other material on this subject, which was found among his effects; he had an active correspondence with other scientists of this field; and he received, with apparent interest, information on the reception of his ballistic studies abroad. One might, therefore, conclude that Ernst Mach did not regard these studies as a minor, casual work.

The motives that might have stimulated Mach to start his gas dynamic studies have occasionally been misinterpreted in the literature[19]. The easiest way to see how Mach became acquainted with this field is to look at the chronological sequence of his publications. His physiological-acoustic experiments led him to perform also physical investigations of acoustic problems, like the reflection and refraction of sound waves. This explains Mach's interest in the blast waves originating from an electric spark discharge, which he regarded at first as acoustic sound waves. After having elucidated the real nature of them as shockwaves (be described below in more detail) he verified the existence of such shockwaves also in connection with projectiles flying at supersonic speeds and in some gas jets from a high-pressure reservoir.

What was Mach's position relative to the problem that his scientific results could have been used for military purposes, i.e., in a way that he might not have wanted? Experimenting with flying bullets was certainly only a means for a definite purpose, the purpose of physical research. In this way one has to understand Mach's saying: "Since, then, shooting and all that appertains thereto, is a very important, if not the most important affair of modern life, you will doubtless not be averse to giving your attention for an hour to some experiments which have been undertaken, not for advancing the ends of war, but for promoting the ends of science, and which throw some light on the phenomena attending the flight of projectiles."

Mach, however, did not disregard the possible consequences of his research:

> No one who has had the opportunity of studying modern guns and projectiles in their marvellous perfection, their power and precision, can help confessing that a high technical and scientific achievement has found its incarnation in these objects. We may surrender ourselves so completely to this impression as to forget for a moment the terrible purposes they serve.[18]

A review of Mach's gas dynamic research makes it necessary to use some technical expressions unfamiliar to the reader who is not acquainted with this field. It appears, therefore, appropriate to explain first a few of these terms which will certainly promote the understanding of Mach's investigations.

The first major group of publications [2–9] concerns the field of unsteady gas dynamics. This refers to a gas stream, the flow pattern of which changes with time: relative to an exterior, fixed observer. Hence the flow field is described by functions that depend on both space and time coordinates. Examples of this type are the air shockwaves or blast waves originated by an explosion. Mach's research was devoted extensively to the study of such shockwaves. (A shockwave is characterized by a discontinuous change of gas pressure as well as density, temperature and other aerodynamic parameters at its front.) In the case of a shockwave originating from an explosion the air in front of the wave is at rest, and the wave moves with supersonic velocity relative to this undisturbed medium. In this way the shock front is a non-linear wave of finite amplitude as analyzed mathematically by B. Riemann [20]. Acoustic waves are a limiting case of shockwaves of vanishingly small amplitude.

When the gas pressure at a fixed position is recorded as a function of time, the pressure is discontinuously increased to a higher level as the shock passes this position. There are two possibilities for the pressure pattern behind the shock front: the pressure either remains constant at that higher level for a certain time, or it re-decreases immediately behind the shock front and eventually reaches a value even lower than the initial (ambient) pressure. The latter takes place when the shock originates from a centered explosion and moves as a spherical wave into the ambient atmosphere. The case of constant pressure behind the shock may occur when the shockwave moves in one-dimensional form, e.g., in a closed, straight pipe. Such a device is called a shock tube. This apparatus consists of a straight tube, usually several meters long and initially separated by a diaphragm into two sections. The gas pressure in one section is increased until the diaphragm bursts; a shockwave will then move into the other section of the tube. One may provide viewing windows in the wall of the tube so that it becomes possible to observe – by means of certain optical devices – the shockwave and its interaction with some models which had been mounted inside the tube. Since the shockwave moves with supersonic velocity one has to apply the methods of high-speed photography in order to get pictures of the unsteady shockwave phenomena. The primary shock is a plane front and appears on the photographs as a straight line perpendicular to the axis of the tube.

When such a plane shockwave hits a vertical wall it undergoes a normal-reflection process, i.e., its direction of motion is reversed. By inclining this wall with respect to the plane of the shock, it becomes evident that the shockwave reflection does not follow the ordinary reflection laws known for the (linear) acoustic waves. If the angle between the wall and the primary shock front exceeds a critical value the reflection pattern exhibits an irregular form (Figures 2a, 2b). This effect was discovered and explained by Mach and was later named after him ('Mach reflection' or 'Mach effect').

Physiological-acoustic investigations led Mach to perform also physical-

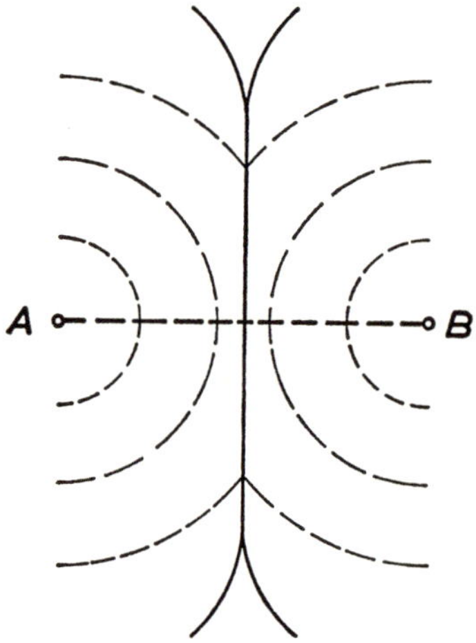

Fig. 1. Principal arrangement for soot-picture experiments.

acoustic experiments. One of his methods to produce sound waves was to generate a weak electric spark discharge[1]. Two years later his assistent Dvorak called his attention to a publication by K. Antolik[21], in which the author reported on "acoustic phenomena of electric spark discharges". This inspired Mach to start with his experimental gas dynamic investigations carried out between 1875 and 1885 at the Physical Laboratories of the University of Prague. They yielded two remarkable results: (1) statements on the nature of shockwaves in air, and (2) the discovery of the irregular reflection pattern of shockwaves.

The principle of these experiments might be explained by means of the arrangement sketched in Figure 1. Two little spark gaps *A* and *B* are ignited simultaneously on the surface of a flat glass-plate covered with a thin sheet of soot. Two spherical blast waves are originated thereby having their centers in *A* and *B*, respectively. While moving over the surface of the plate the traces of these waves blow off a big portion of the soot. However the two blast waves strike eachother and interfere on a line halfway between *A* and *B*; their mechanical action is thereby canceled; the soot remains untouched on this interference line which is, therefore, well marked. At a certain distance from the connecting line $\overline{AB}$ this interference curve separates into two parts forming the two branches of a '*V*'; this is the onset of the Mach effect. The two branches of the '*V*' are the traces of the triple-point on the plate.

If the spark gap is chosen to be not a point but of finite length, then the resulting blast waves are not spherical but cylindrical, and their trace on the sooted plate appears as a plane wave front. By varying the shape of the electrical discharge one may obtain rather complicated, symmetrical or non-symmetrical soot pictures. Antolik's experiments were no more than playing with such complicated soot pictures, see e.g. Figure 3. However Antolik, a high-school teacher in Hungary, speculated that one could learn something from these pictures about the characteristics of electrical discharge.

Mach recognized soon that the formation of the soot pictures was due only to mechanical-acoustic effects of the spark discharge and that it was no electric phenomenon. This conclusion was confirmed when he generated the same soot pictures by little explosions, i.e., by non-electric processes. The physicist August Toepler had already photographed spherical and cylindrical blast waves (Figure 4). At first it was Mach's opinion that these waves originating from the spark discharge were simple acoustic waves. He was able, however, to correct this mistake after having measured their velocity. But these measurements had initially a quite different aim: since Mach regarded the waves to be acoustic and since the velocity of sound was very well known, he thought that the formation of the interference line in the soot could be used for measuring little time intervals in a highly exact way. If one of the spark gaps is fired a little later than the other the interference line will be shifted towards the spark ignited with some delay. This shift divided by the velocity of sound should yield the delay time between firing the two gaps. These time intervals were expected – according to the dimensions in Mach's experiments – to be less than one millisecond.

Mach's apparatus for delaying the ignition of the two sparks was to release the discharge by means of a flying bullet. The time required by the bullet to fly from one spark to the next is the delay time; the speed of the bullet could be very exactly determined with a ballistic pendulum. These measurements showed, that the velocity of the blast waves was above the regular velocity of sound, that it increased with the intensity of the discharge (or the explosion), and that it decreased with increasing distance from the center of origin. At a great distance from the source of explosion the blast wave velocity approached the value of the speed of sound. A well-refined technique resulted in very exact velocity diagrams of the

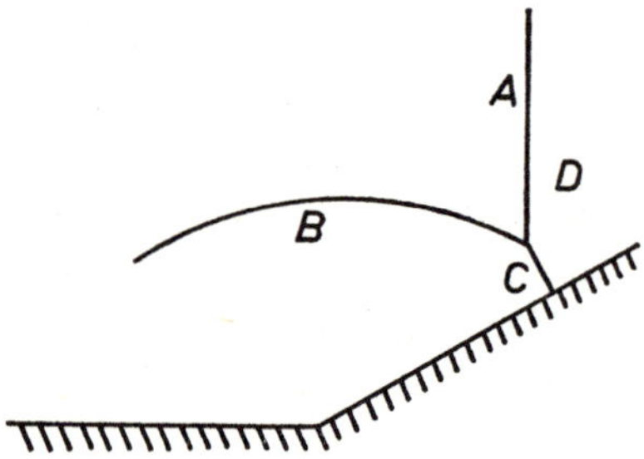

Fig. 2a. A = primary shockwave, B = reflected shockwave, C = 'Mach stem', D = triple point.

blast-wave motion obtained by Mach, Tumlirz and Kögler[4] (Figure 5). Hence the blast waves, at least in their initial phase, were not identical with the acoustic waves of infinitesimally small amplitude; Mach concluded that they were rather shockwaves of finite amplitude as already treated theoretically by Riemann[20]. Mach writes, "It does not contradict the theory to assume that the velocity of sound increases with the intensity of the impulse. Only for very small vibrations does the velocity of sound not depend on the amplitude. But this is not valid for vibrations of finite amplitude as has been proved by Riemann in his paper 'Uber die Fortpflanzung ebener Luftwellen von endlicher Schwingungsweite', 1860. 'Velocity of sound' receives hereby a quite different meaning; it is different at every point of the wave and alters during the wave motion. It appears that we deal in our experiments with such waves as described by Rie-

Fig. 2b. Shadowgraph displaying the Mach reflection of an unsteady shockwave at an inclined wall. (The primary shock front moves from the left to the right.)

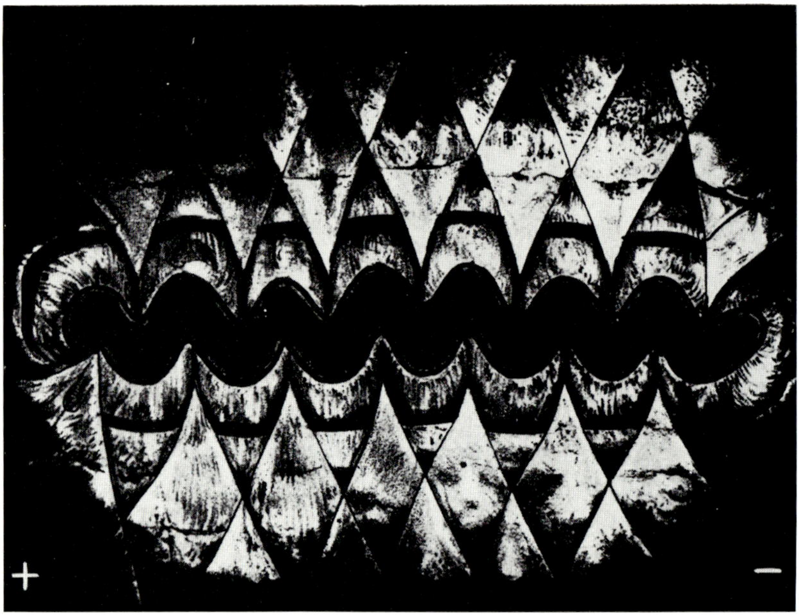

Fig. 3. Soot picture obtained. (K. Antolik[21])

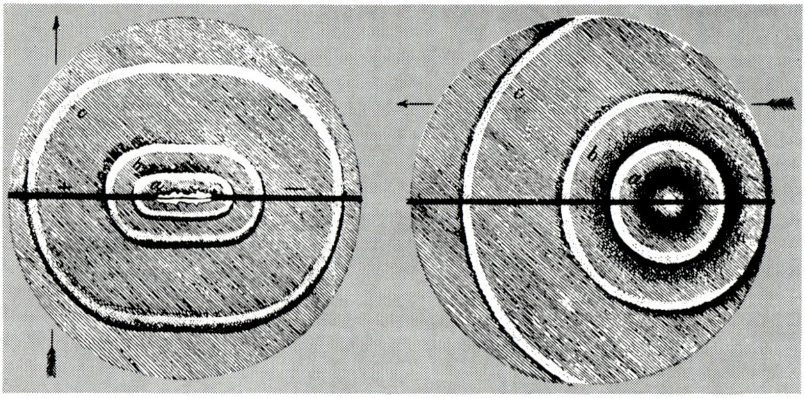

Fig. 4. Toepler's schlieren photographs of spherical and cylindrical blast waves originating from a pulsed spark gap.

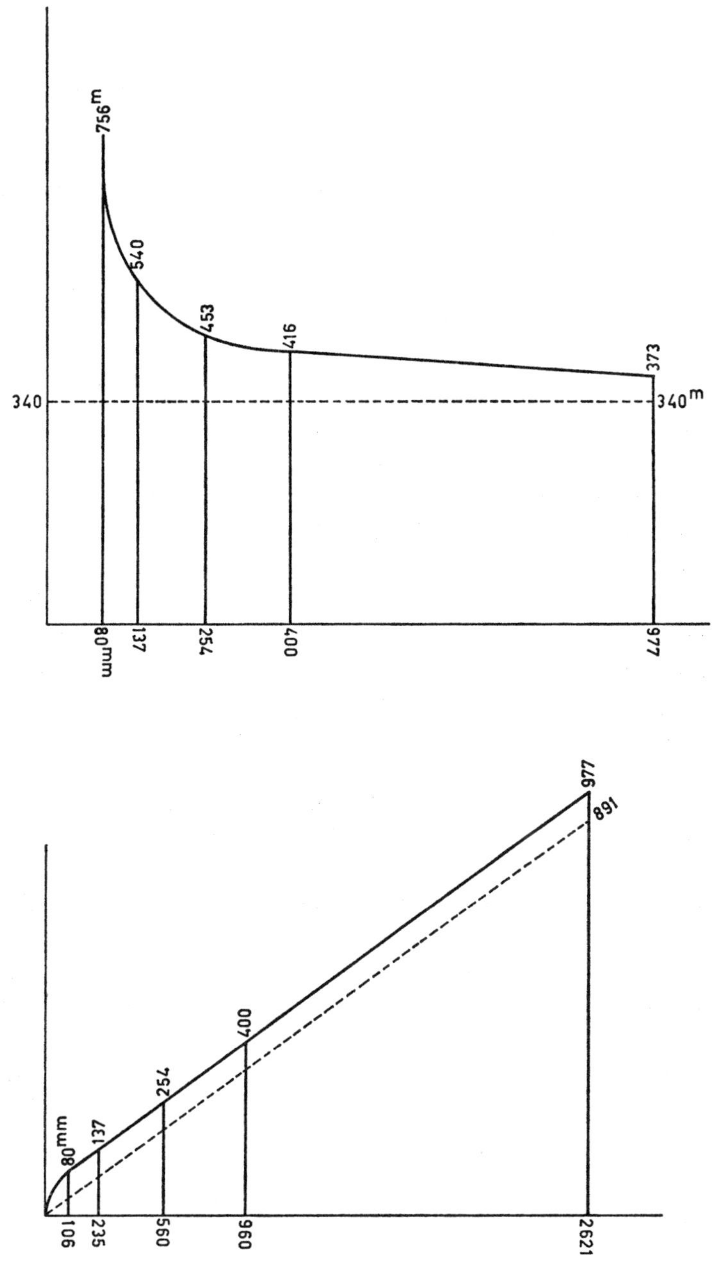

Fig. 5. Velocity diagrams of the blast-wave motion obtained by Mach *et al.*[4] (The blast wave velocity approaches the acoustic speed at greater distance from its center of origin.)

mann."[3] Mach still uses here 'velocity of sound' for the speed of the blast waves.

According to their original aim the soot-picture experiments had yielded a negative result: neither did they give an insight into the process of electrical discharge, nor did they allow for an exact measuring of short time intervals. The unexpected, positive result was the discovery of shock-waves in air. It seems significant for Mach's thinking that he did not rest content with having achieved the 'indirect' visualization by means of the sootpictures; he rather wanted to realize a 'direct' visualization in an optical way. Toepler's schlieren method appeared appropriate for this purpose. Using this method Mach photographed the two-dimensional shockwave formation between two parallel glass plates, an arrangement which is a precursor to the modern shock tubes. Later he used Jamin's interferometer in connection with a photographic device for quantitative measurements of the shock strength[7]. With this method – from which he and his son Ludwig developed years later the well known Mach-Zehnder-interferometer – he found out that the relative change in density at the shock front is about 50 times what had been measured for regular sound waves, e.g., by himself as well as by Toepler and Boltzmann (Figure 6). Mach mentions, however, that these measurements do not yet allow one to prove Riemann's theory, which we know today was not wholly correct in its thermodynamic assumptions. "Since each blast wave is an individual, this theory can only be re-examined when one is able to photograph the same wave in two succeeding phases. It appears to be possible that this aim can be achieved"[7]. It is not known whether Mach had found already such a device comparable to a multiple-spark camera.

Though Mach had given precise information on the nature of shock-waves there were still some strange theories or pseudotheories described in the literature, especially on explosion waves, e.g., a theory by Abel to which Mach refers in[9]. An exploding charge, e.g., a dynamite cartridge, can cause a second charge at some distance from the first to explode also. According to Abel's theory the explosion wave was supposed to be a periodic wave which would cause the second load to vibrate and, due to a certain resonance effect, to explode. Mach refuses this interpretation in the paper[9] of 1885, and he emphasizes again the characteristics of shockwaves; namely, they consist of one shockfront without any periodicity and can cause mechanical forces like a flying bullet. In this paper he reports also on measurements made to determine the speed of the explosive front in fulminating mercury. Using the above-mentioned shift of interference lines on a sooted plate he measured speeds up to 2000 meters per second.

The 'V'-shaped separation of the interference lines on the sooted plate was correctly interpreted by Mach[5] in 1878. He had noticed after each

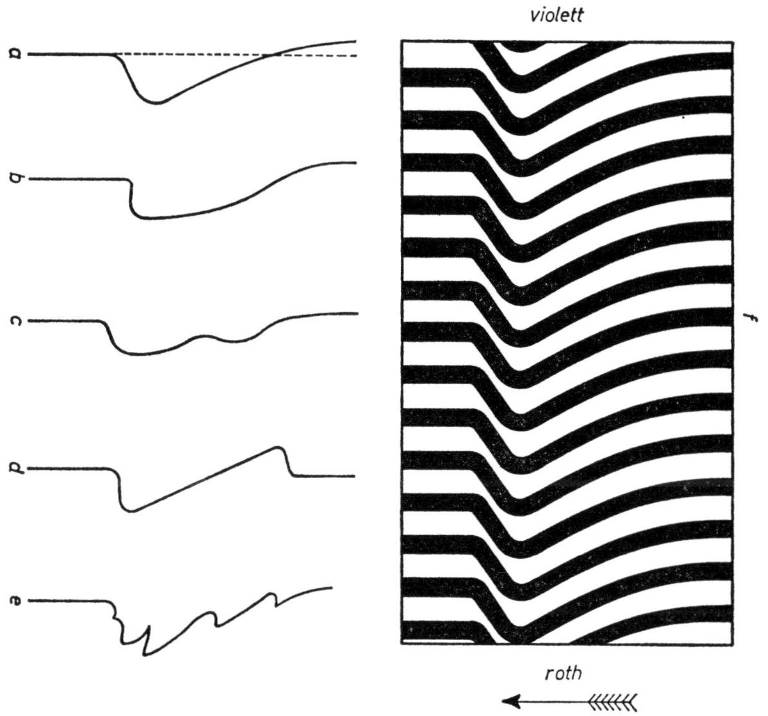

Fig. 6. Interferogram displaying the density jump at the front of a shockwave.[7]

experiment that the soot inside the two branches of the 'V' was more heavily compressed than outside the 'V'. The air therefore, above the V was apparently compressed by a stronger wave than the air outside of this zone. Two interference effects instead of one must have occurred within the V-zone.

Linear waves, i.e., waves of vanishingly small amplitude, may interfere without reacting with each other. Initially this appears to be the case also for blast waves, namely, where only one interference line is drawn in the soot. According to Mach the intersection of two waves of finite amplitude becomes the source of a new wave which travels along the primary wave front with a certain velocity ω. When the velocity component of the secondary wave, parallel to the direction of motion of the primary wave, $\omega \cdot \cos(\varphi/2)$, exceeds the speed W of the primary wave, then the V-shaped separation is supposed to start, and the irregular reflection begins. It is possible to determine for given W and ω a limiting angle φ_0 below which regular reflection cannot exist. W, ω and hence also φ_0 are functions of the shock strength. The phenomenon is the same when the axis in Figure 1 is replaced by a solid wall, at which the blast wave originating from one spark gap is now reflected.

The analytic investigation of this type of reflection was only started during world war II. (It was R. J. Seeger who named the irregular reflection of shockwaves a 'Mach effect' or 'Mach reflection'.[23]) Mach's concept of the 'wave on the wave front' is also used in the very modern analysis of this problem by G. B. Whitham[24]. It should be noted that the interpretation of the irregular reflection pattern by Mach, whose ideas have been reported here, was based only on the observation of the soot pictures and not on an optical visualization of this effect.

The name of Ernst Mach has become most famous in the field of aerodynamics due to his ballistic research. With these experiments, which were started in 1886, he discovered the essential characteristics of supersonic flows. But Mach was not unprepared when he started these investigations. During his studies on unsteady shockwaves he had already recognized the importance of the velocity of sound for the motion of a gas. In the paper[9] of 1885 he wrote: "Only those cases are familiar to our experience and to our mechanical instinct, in which the velocity of motions and deformations is small compared to the velocity of sound. In the opposite case very surprising phenomena appear which are unknown to our intuition. They lead to a new class of mechanical problems and still need a theoretical analysis." Hence the way Mach was to draw his conclusions from ballistic experiments was inevitably indicated to him. Today it appears possible that Mach had already a certain conception, e.g., on the existence of a bow shock, before the experiments were started, and that these investigations only confirmed Mach's scientific concept.

In the same paper[9] Mach uses the ratio of flow speed to acoustic speed as a special parameter. The expression 'Mach number' for this ratio was introduced by J. Ackeret[25] in 1929.

It is interesting to be aware of the reasons motivating Mach to start the ballistic experiments. One reason was the well-known observation of artillerymen that one perceives usually two bangs while a fast projectile is flying by. One of them had to be the bang originating at the muzzle of the gun. The second, however, found no appropriate explanation; it did not appear when the speed of the projectile was rather low. The second reason can be traced back to the German-French war of 1870/71. It had then been noticed that the French Chassepôt-bullets caused big, crater-shaped injuries; the French had been charged, therefore, with using explosive projectiles and thus violating the international treaty of Petersburg (1868), which prohibited the use of such projectiles. These charges had been disproved by the Belgian physicist M. Melsens who reported[26] on his investigations in 1872 at the Royal Academy of Sciences in Brussels. He made the following suppositions: (1) a spherical projectile is able to carry a considerable amount of compressed air which depends on its velocity; (2) the compressed air precedes the projectile and may cause mechanical, explosion-like effects. The reported injuries have to be explained in this way.

It was Mach's aim to prove this hypothesis by Melsens. The way to perform such experiments was well designated since the necessary technical apparatus was already known. The flying projectile and the phenomena in the ambient air had to be photographed. The pressure disturbances in air could be made visible with Toepler's schlieren method. In order to focus the fast flying projectile an electric spark of short duration provided the photographic exposure. Such an arrangement (Figure 7) had already been used by Toepler and Mach. Before entering the field of view the projectile itself triggers the spark light-source. The optical set-up has to be placed in a closed, darkened room.

P. Salcher, a professor at the Naval-Academy in Fiume (today: Rijeka, Yugoslavia), actually performed the experiments upon invitation by Mach, who provided a certain experimental program. The discussion between Salcher and Mach, who still lived in Prague, took place in the form of a very frequent exchange of letters. Salcher reported on the progress of his investigations about every week or even more often. A collection of 140 letters of Salcher addressed to Ernst Mach is available at the Ernst-Mach Institut in Freiburg (Germany), the first of them dated February 14, 1886. The letters of Mach, who had apparently written just as frequently, have unfortunately not been preserved. From the extant letters it is reconstructable that Mach informed Salcher as to the correct interpretation of the new phenomena shortly after having received the reports, sketches and photographs of the experiments. It is not surprising that Mach was able to give these explanations immediately. If one considers the results of the studies on shockwaves it becomes obvious how familiar Mach already was with these new problems.

The first ballistic experiments were performed in Fiume at bullet speeds between 400 and 500 meters/second and calibers of 8 mm. The results[10] were published in 1887. The first photographs which Salcher made of bullets flying at supersonic speed displayed the characteristic bow shockwave (Figure 8). Mach interpreted this wave at once as the envelope of disturbances originating from the projectile, and he supposed it to be a

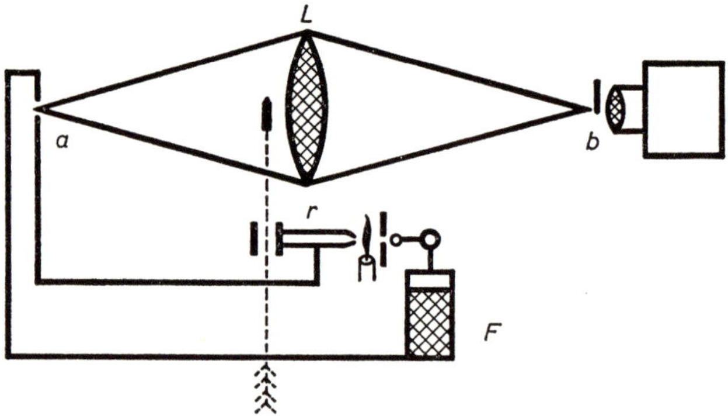

Fig. 7. Photographic arrangement using a spark source for recording a flying bullet.[18]

shockwave. So it was possible to solve the two old ballistic problems, the appearance of two bangs and the explosive-like injuries: the first bang is the shockwave carried along by the projectile; the second is the muzzle bang which moves with sound speed and is therefore slower than the projectile. The explosion-like injuries were caused by the increased pressure behind the bow shock. Melsens' hypothesis that the projectile carried a certain amount of compressed air attached to the bullet, was thereby disproved. The correct explanation by Mach was that the compression of air by a supersonic projectile is a dynamic effect and not a stationary one.

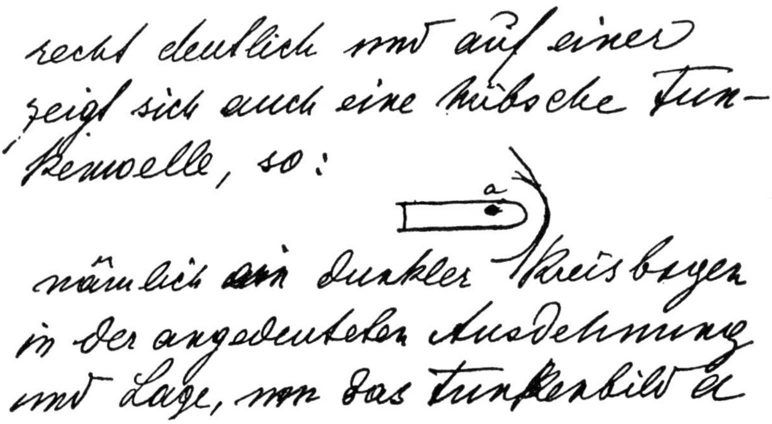

Fig. 8. Sketch of observation of a bow shock of a bullet at supersonic speed. (Part of a letter of P. Salcher to Ernst Mach, dated May 23, 1886.)

In December 1886 Mach formulated – based on photographic observation – the law for the 'Mach angle' α, named after him:

$$\sin \alpha = \frac{\text{acoustic speed}}{\text{flow speed}}.$$

Also the 'Mach waves' arising from the rough surface of the projectile and the wake pattern behind the body found an immediate interpretation by Mach. All these were phenomena which Salcher observed first but which he found perplexing and at first unexplainable. In a letter of December 22, 1886, he wrote to Ernst Mach: "The available literature and my experience are not sufficient to explain the phenomena in question."

Using higher flight speeds and larger projectiles these ballistic studies were later continued by Salcher in Pola (Yugoslavia) and by Ernst Mach and his son Ludwig in Meppen (Germany) with support from the Krupp-company.[12,13,15,16] The quality of the photographs was refined; more details became visible; the new experiments confirmed the first results by Mach and Salcher; in principle, however, they brought nothing new. The pressure change at the bow shockwave could be measured quantitatively when the interferometer designed by Ernst and Ludwig Mach (later 'Mach-Zehnder-interferometer') was applied instead of the schlieren system (Figure 9).

It should be mentioned that Mach was able to explain another phenomenon related to his ballistic findings and which some scientists had reported previously in the literature. A meteor falling down to the surface of the earth produces first a sharp bang, and only thereafter one hears the noise of the impact. This bang is nothing else than the bow shock reported by Mach, since the meteor comes down through the atmosphere with supersonic velocity[17].

Rather by chance Mach was led to the observation of a further gasdynamic phenomenon, the formation of certain wave patterns in a free-air jet exhausting from a pressure reservoir through a small opening. The experiments were again performed by Salcher in a torpedo factory in Fiume; Mach and Salcher published the results[14] in 1889. These experiments are interesting in so far as this was certainly the first time that the idea of a supersonic blow-down wind-tunnel had been expressed. In this paper the authors say: "On occasion of the experiments with flying projectiles Salcher had the idea to investigate experimentally also the reverse case with the air flowing and the test body being at rest, in order to prove the obtained results"; and in a footnote, "If the jet has a big enough cross-section, a phenomenon analogous to the bow shock becomes visible at the fixed body." This remark proves that supersonic velocities did really exist in the exhausting jet, a fact, which was later denied (see below).

Another phenomenon, however, attracted the attention of the experimentalists. The air jet was photographed by Salcher by means of a schlieren system. The pictures displayed periodical crossed lines within the jet which Salcher used to call 'Lyra', when he described this observation to Mach (Figure 10). At first Mach explained these lines as stationary acoustic waves, but recognized later that some of these lines had to be compression waves or shockwaves analogous to the bow shock of a projectile.

Despite Mach's interpretation the wave phenomena in a free jet were later misunderstood, since the terms 'compression waves' and 'acoustic

Fig. 9. Interferogram of a bullet. (Mach[18])

waves' were confused. Some authors regarded the crossed lines as sound waves and concluded that the jet had sonic velocity and that the velocity of a free jet could not exceed the speed of sound. It was L. Prandtl[27] who gave the analytic treatment of the wave phenomena in 1904. Expan-

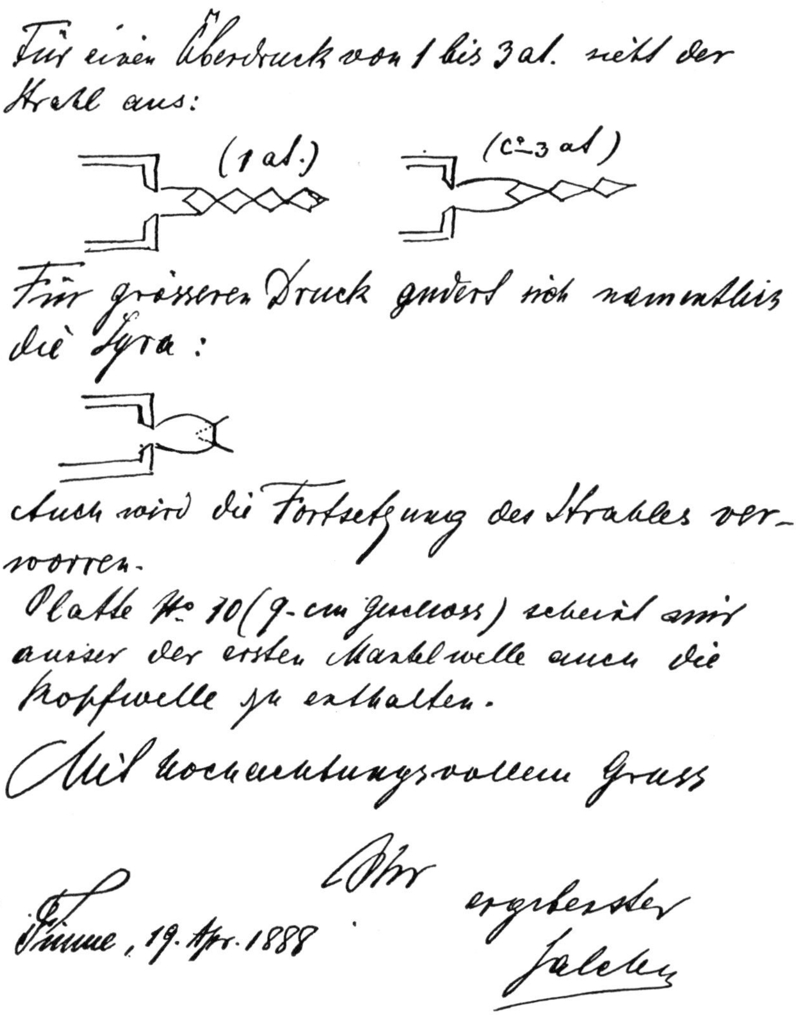
Für einen Überdruck von 1 bis 3 at. sieht der Strahl aus:

(1 at.) (c. 3 at)

Für grösseren Druck ändert sich namentlich die Spitze:

Auch wird die Fortsetzung des Strahles verworren.

Platte No 10 (9-cm Geschoss) scheint mir ausser der ersten Mantelwelle auch die Kopfwelle zu enthalten.

Mit hochachtungsvollem Gruss

Ihr ergebenster
Salcher

Fiume, 19. Apr. 1888

Fig. 10. Wave phenomena in a free-air jet. (Part of a letter of P. Salcher to Ernst Mach, dated April 19, 1888)

sion waves originate at the edge of the outlet; they are reflected at the boundary of the free jet as compression waves, which in turn are again reflected as expansion waves. This process repeats periodically.

The dominating phenomenon in Mach's gas-dynamic research was the shockwave. Mach's methodic approach to this field necessarily led to the discovery of shockwaves, for the optical methods he applied were most sensitive to the visualization of shocks in air. Mach recognized properly the importance of the acoustic speed for these phenomena. He made accessible a new field in mechanics in which the speed of motions is of the same order of magnitude as the velocity of sound. His experiments opened to view the problems of supersonic aerodynamics, and Mach's findings determined the trend in this field for several decades.

Mach's optical arrangements have been continuously improved and refined since then. While his methodology found little success in other fields of physics and while many physicists turned away from Ernst Mach, the aerodynamicists always remained pure Machian. The optical visualization of compressible flows is still regarded as one of the most fruitful experimental methods. So it is not surprising that the name of Ernst Mach has been chosen by a research institute which is deeply involved in applying optical visualization methods for its experimental research. In this sense the use of Ernst Mach's name is entirely appropriate, not controversial as in some fields of physics and philosophy; his name stands here for the indisputable success of Mach's methodical thinking in this field.

Ernst-Mach-Institut
Freiburg i. Br.

REFERENCES

* The author gratefully acknowledges the assistance of Dr. R. J. Arenz in translating this manuscript.

[1] E. Mach and A. Fischer, *Sitzungsberichte der kaiserlichen Akademie der Wissenschaften. Mathematisch-naturwissenschaftliche Classe. Wien.* (=*SW*) **67** (1873), 81–86.
[2] E. Mach and J. Wosyka, *SW* **72** (1875), 197–212.
[3] E. Mach and J. Sommer, *SW* **75** (1877), 101–130.
[4] E. Mach, O. Tumlirz, and C. Kögler, *SW* **77** (1878), 7–32.
[5] E. Mach, *SW* **77** (1878), 819–838.
[6] E. Mach and G. Gruss, *SW* **78** (1878), 476–480.
[7] E. Mach and J. von Weltrubsky, *SW* **78** (1878), 551–560.

[8] E. Mach and J. Simonides, *SW* **80** (1879), 476–486.
[9] E. Mach and J. Wentzel, *SW* **92** (1885), 625–638.
[10] E. Mach and P. Salcher, *SW* **95** (1887), 764–778.
[11] E. Mach, *SW* **97** (1888), 1045–1052.
[12] E. Mach and P. Salcher, *SW* **98** (1889), 41–50.
[13] E. Mach, *SW* **98** (1889), 1257–1276.
[14] E. Mach and P. Salcher, *SW* **98** (1889), 1303–1309.
[15] E. Mach and L. Mach, *SW* **98** (1889), 1310–1326.
[16] E. Mach, *SW* **101** (1892), 977–983.
[17] E. Mach and B. Doss, *SW* **102** (1893), 248–252.
[18] E. Mach, 'Über Erscheinungen an fliegenden Projektilen', *Populär-wissenschaftliche Vorlesungen*, (3rd ed.), Leipzig 1903, pp. 351–377. (Excerpt from translation by T. J. McCormack, 5th ed., Open Court, La Salle, Ill., 1943.)
[19] H. Henning, *Ernst Mach*, Leipzig, 1915.
[20] B. Riemann, 'Über die Fortpflanzung ebener Luftwellen von endlicher Schwingungsweite', *Abhandlungen der Göttinger Gesellschaft der Wissenschaften*, 1860.
[21] K. Antolik, 'Das Gleiten elektrischer Funken', *Poggendorfs Annalen der Physik und Chemie* (1875), 14–37.
[22] A. Toepler, *Beobachtungen nach einer neuen optischen Methode*, Bonn, 1864.
[23] P. C. Keenan and R. J. Seeger, 'Analysis of Intersections of Shockwaves', US Navy Bu Ord *Explosives Research Report* 15, 1944.
[24] G. B. Whitham, 'A New Approach to Problems of Shock Dynamics', *J. Fluid Mech.* **2** (1957), 145–171.
[25] J. Ackeret, *Schweizerische Bauzeitung* **94** (1929).
[26] M. Melsens, 'Sur les plaies produites par les armes à feu', *Société Royale des Sciences Médicales*, Brussels 1872.
[27] L. Prandtl, 'Über stationäre Wellen in einem Gasstrahl', *Physikalische Zeitschrift.* **5** (1904), 599–601.

Note: Figure 2b has been made available by courtesy of the late Prof. Hubert Schardin, Institut Franco-Allemand de Recherches, Saint-Louis, France.

RAYMOND J. SEEGER

ON MACH'S CURIOSITY ABOUT SHOCKWAVES

Ernst Mach was basically an experimental physicist. Even upon his appointment as professor of the History and Theory of Inductive Science at the University of Vienna (1895) he announced in his inaugural lecture, "I shall remain a natural inquirer."[1a] Ten years later, in the Preface to *Erkenntnis und Irrtum*, he claimed, "I am not a philosopher at all, but only a scientist."[2] At the very beginning of his professional career (1865), indeed, in a popular lecture at Graz, he had already set forth the creed of a serious inquirer: "Whoever has roamed through a beautiful country knows that the tourist's delights increase with his progress. It is also true of the natural philosopher ... the first questions are formed upon the intention by the inquirer by practical considerations; the subsequent ones are not. An irresistible attraction draws him to these; a nobler interest which far transcends the mere needs of life."[1b]

Mach, however, was a humanistic physicist with a historical perspective and a philosophical view. As he emphasized in his inaugural address as Rector of the University of Prague (1883), "The factor which most promotes scientific thought is the gradual widening of the field of experience."[1c] In *The Analysis of Sensations* he made his meaning explicit by indicating the reason; in the *Introduction* he remarked, "I do not claim the title of philosopher. ... I want only to take, in physics, a standpoint which does not have to be abandoned immediately when we look over into the field of another science. For all the sciences ultimately form a whole."[3] In short, he wished to employ only those concepts of physics which would not lose their usefulness outside of physics.

His broad humanistic interests culminated in a zealous desire to communicate. In the Preface to *Popular Scientific Lectures* he argued, "By an appropriate choice of the matter, the charm and the poetry of research can be conveyed... It is only necessary to set forth the attractive, the alluring features of a problem, and to show what broad domains of fact can be illuminated by the light radiating from the solution of a single and ofttimes unobtrusive point."[1] Despite Mach's own disparaging comment

Boston Studies in the Philosophy of Science, VI.

about "the dry domain of physics"[1b] these lectures are still quite enjoyable. As noted by Carlton Berenda Weinberg, "Mach, himself, probably arrived at the principle of economy by way of his experience as a pedagogue."[4] In a Vienna lecture (1882) he himself explained that "the aim of instruction is simply the saving of experience."[1d] He concluded, "Physics is experience arranged in economical order."

Mach's interests as a humanistic physicist are indicated in the distribution of subjects among his published popular lectures:[1] five are on physics per se, four on sensation, and five on scientific thought. Note, however, that these lectures are interrelated and that all are based on physics. Because of my own interest in the physics of fluids, I must confess some disappointment in noting in his celebrated *Mechanics*[5] that only 32% of the 149 pages allotted to statics are on hydrostatics, and that only 10% of the 365 pages on dynamics are strictly hydrodynamics. His interest in fluid dynamics, to be sure, was focused mainly on wave propagation in compressible fluids. Accordingly, in the third edition (1898) of the popular lectures we find one additional lecture, namely, "On Some Phenomena Attending the Flight of Projectiles"[1e] (given that very year).

In this lecture, he referred to one he had attended in Paris by the Belgian artillerist M. Melsens. He recollected, "The desire arose in me of experimentally testing his conjecture and of rendering the phenomenon [shockwave], if it really existed, perceptible ... The desire was the stronger as I could say that all the means for realizing it existed, and that I had in part already used and tested them for other purposes."[1e] Mach was referring to the schlieren technique developed in 1864 by August Joseph Ignaz Töpler (1836–1912) and used successfully by him in 1867 to make sound waves visible by means of a spark. In his own experiments in 1884 Mach employed a target pistol. He failed, however, to find the expected headwave. Later he realized that the projectile had a speed of only 240 meters per second, considerably less than that of sound. Experiments conducted on an open artillery range with higher speed projectiles remedied this defect. The laboratory, however, attracted him more for investigations because of its potentially greater precision – not to mention the considerably less expense. The photograph[1e] shown below (Figure 1) is that of a blunt projectile taken with an Austrian Mannlicher rifle.

An interferogram was the experiential basis for Mach's conclusion that a headwave is relatively thin. He explained this headwave by applying

Christian Huygens' (1629–1695) principle for acoustic waves to a disturbance produced by a moving projectile; he related the angle of inclination of the envelope of the secondary wavelets to the ratio of the projectile speed to the acoustic speed – the so-called Mach number.

Mach conducted further experiments by having a spark from a Leyden jar discharge through a V-shaped metallized thread on a glass plate covered by another parallel plate a few millimeters distant, the lower surface of the upper plate being coated with soot. Figure 2 (1875) shows the funnel-shaped area swept clean of soot upon discharge.[6] Wood[9] enthusiastically remarked about the 'ingenious method' by which Mach was thus able to 'photograph' an instantaneous wave by having another wave propagated to meet the initial disturbance produced by the V-shaped thread. Although a photograph of this particular experiment is not available, a similar effect (Figure 3) is shown in the wave pattern produced by wave fronts from two Mach V's opposite each other.[6]

Concluding his lecture on the photography of projectiles Mach remarked, "Researches of the kind we are considering can certainly not be detrimental to practical artillery. It is no less certain that experiments by artillerists on a large scale will be of undoubted benefit to physics."[1e] This prophecy was borne out in World War II.

Colonel Hermann H. Zornig (1888) of the Watertown Arsenal had called the attention of John von Neumann (1903–1957) to Mach's soot experiments. The latter speculated that they might be related to the Charles Edward Munroe (1849–1938) effect for shaped charges. In 1949 Edgar Bright Wilson, Jr., (1908–) with the aid of W. D. Kennedy[7] and J. Frankl repeated some of these soot experiments. Von Neumann[8] proposed a simple theory in his fundamental report on the 'Oblique Reflection of Shocks'. His primary concern was to find, if possible, an equivalent Huygens principle for shockwaves, which have a finite amplitude. For further investigations, he was fortunate to enlist the experimental curiosity of Robert Andrews Wood[9] (1868–1955), who had himself employed Töpler's method at the turn of the century to photograph acoustical phenomena, and whom Eduard da Costa Andrade (1887–) has called "one of the greatest experimenters of physics of his time". In 1943, Wood, then research professor (retired) at The Johns Hopkins University, began by making some improvements in the original Mach method. Specifically, he drew lead-pencil lines on finely ground glass and thus obtained greater consistency than had been possible with merely aluminized threads.

The parabolic interaction of circular and linear waves had already been noted by Mach; at the extreme ends (larger angles), however, he had found swept-out funnels. This phenomenon, for two circular waves, is illustrated in Figure 4. Wood himself made more than 50 soot plates of which two typical ones are shown in Figure 5. The halo surrounding the original discharge in each instance had been explained by Mach as the result of reflections between the plates (cf. the effect of a virtual image). Wood was able to realize also the so-called Mach bridge (cf. Figure 6.), which had not been obtained by Mach. Actually Wood's experimentation involved many unfinished investigations; for example, the existence of double lines, both for the bridge and for the other wave, as seen under different magnifications in Figure 7. These pictures show also the so-called line of discontinuity (same pressure, but different densities), which had been noted first in spark photography by Alexander Crane Charters (1913–) then at the Ballistic Research Laboratories in Aberdeen, Md.

Mach had also applied schlieren techniques to study shockwaves; indeed, his setup became the classical one used for illustration in Karl Julius Cranz's (1858–1945) well-known *Lehrbuch der Ballistik*. Accordingly, Wood investigated interactions of shockwaves with the schlieren method. He used essentially the same electrical equipment (Wood, however, employed a Leyden jar charged from a 6-inch gap of an induction coil), but added a photographic plate. No cover plate, of course, was used in this case. Figure 8a and 8b show the growth of the Mach bridge for different times of photographing; Figure 8c shows the discontinuity. As George B. Kistiakowsky (1900–) noted in the Preface to Wood's report, "There was general confirmation of von Neumann's idea."[9]

The interaction of shockwaves was involved in still another investigation of Ernst Mach and his collaborators, particularly his son Ludwig. If a gas receiver has a pressure less than a certain critical pressure in a connecting reservoir (0.527 for a perfect diatomic gas), the issuing supersonic jet exhibits a schlieren pattern of intersecting shockwaves – with a definite spatial periodicity (identified in 1889). In 1897 Ludwig Mach (1868–1951) showed by both schlieren and interferometric techniques the Mach pattern in Figure 9, as well as the regular X pattern. (It is noteworthy that the so-called Mach-Zehnder interferometer developed in the 1890's was used also in World War II for similar investigations).

Rudolf Walter Ladenburg (1882–1952) repeated these experiments[10] and analyzed them on a quantitative basis. Here, too, observations agreed generally with simple theoretical considerations. The relationship of the Mach pattern of stationary shockwaves to those of interacting progressive waves had apparently not been noted by Mach. It was von Neumann, however, who visualized the same theoretical explanation for both these phenomena.

From an examination of Mach's own investigations along this line one can discover clues for understanding his scientific methodology. One finds that he preached what he practised. We are confronted at once with his reliance upon sense experience. As he had remarked in 1867, "Not without reason ... did Lord Bacon call the experimental method a questioning of nature."[1f] And thirty years later he emphasized that "modern science strives to construct its picture of the world not from speculations but so far as possible from facts."[1e] His basic concern for empirical means of obtaining knowledge developed into an antimetaphysical attitude. He proclaimed in the *Preface* to his *Mechanics* (1st edition, 1883), "The present volume is not a treatise upon the application of the principles of mechanics. Its aim is to clear up ideas, expose the real significance of the matter, and get rid of metaphysical obscurities."[5] With respect to so-called "humanistic sciences" he argued, "Can it really be that higher culture is to be gotten only from a few old spots and from palimpsests which are at best mere scraps of nature, or that more is to be learned from them alone than from all the rest of nature?"[1g] Still later he made the sobering remark. "All our art would be in vain did not nature herself afford at least some guiding threads from a hidden phenomenon into the domain of the observable."[1e]

Mach himself, moreover, used existing laboratory techniques whenever possible. Starting with Huygens' use of the telescope for detecting striae in glass and Jean Léon Foucault's (1819–1868) intercepting slide (the so-called schlieren method), Töpler had combined these to make visible sound waves from electric sparks. Similarly, Mach used the flying projectile itself to trigger an electric spark and thus to obtain a picture.

Finally, what was Mach's concept of the role of theory in such investigations? In the first place, he believed in the critical character of exploratory research. In his very photography of projectiles, he mused, "It is always peculiarly fascinating ... to subject to direct verification by obser-

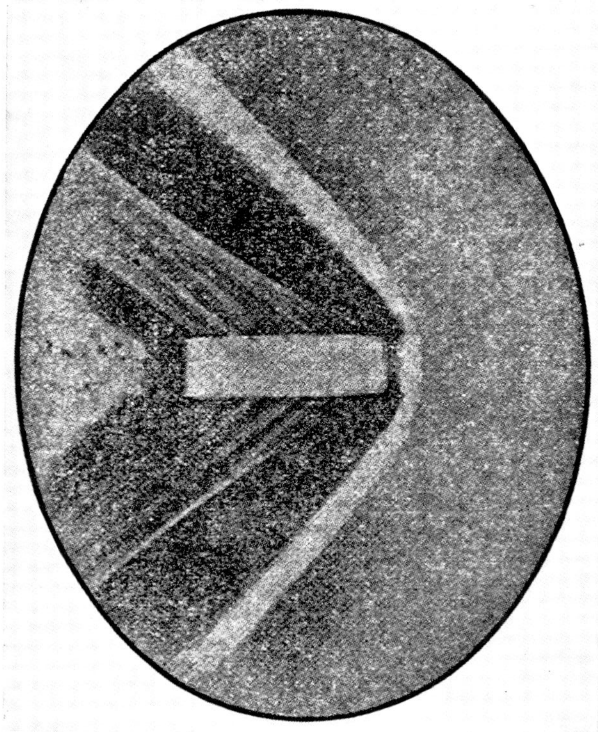

Fig. 1. Schlieren picture of a supersonic blunt projectile. (E. Mach)

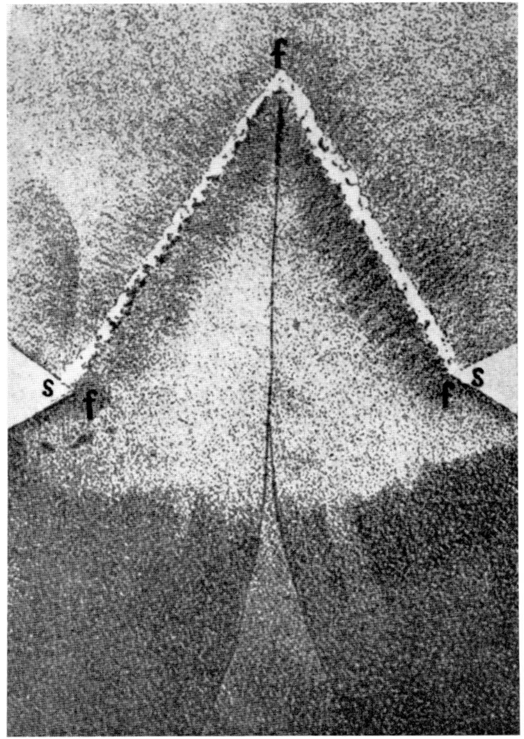

Fig. 2. Soot funnel from a V-spark. (E. Mach)

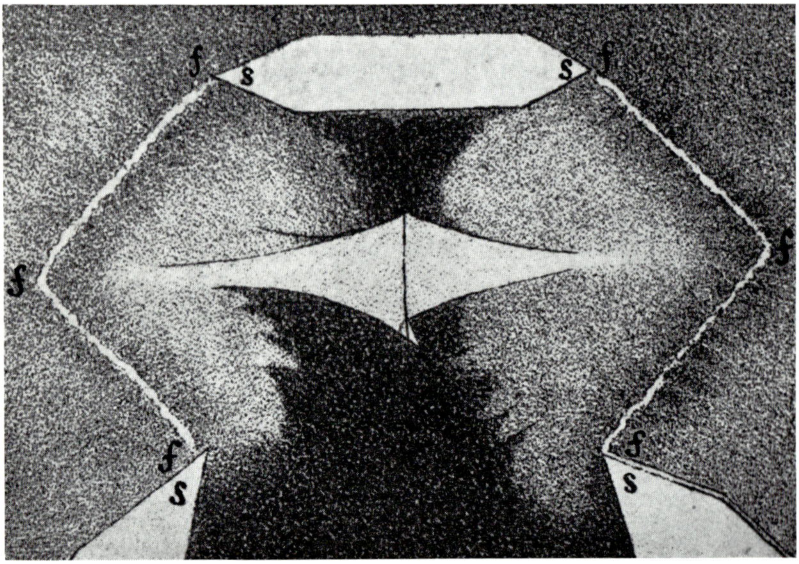

Fig. 3. Soot funnels from two opposing V-sparks. (E. Mach)

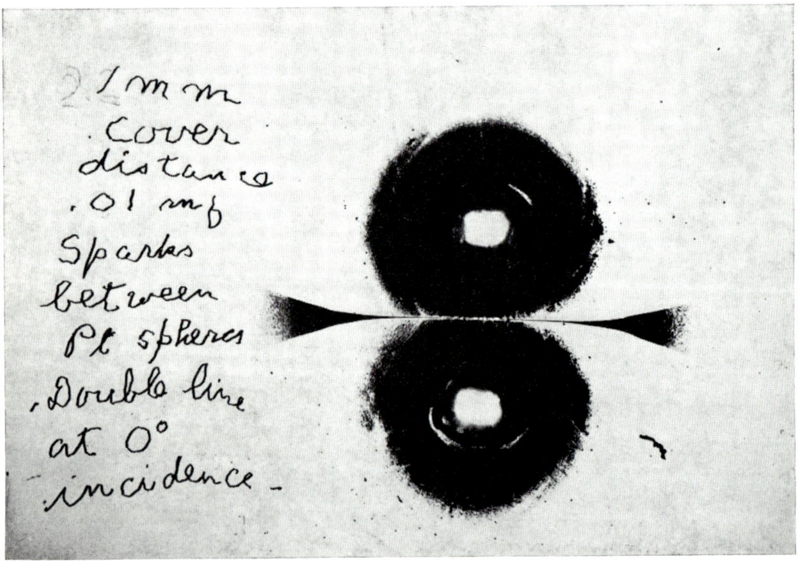

Fig. 4. Soot funnels for two interacting 'circular waves'. (R. W. Wood)

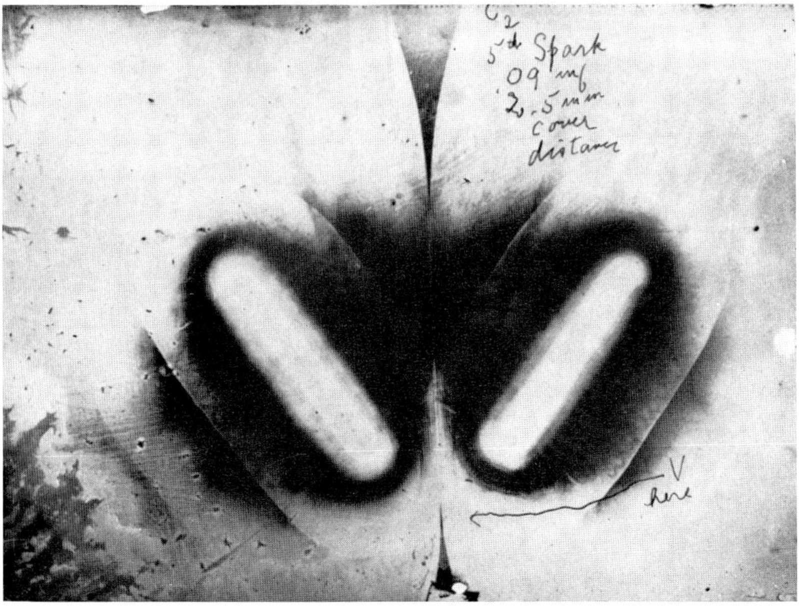

Fig. 5a. Soot funnel for Mach V-narrow angle. (R. W. Wood)

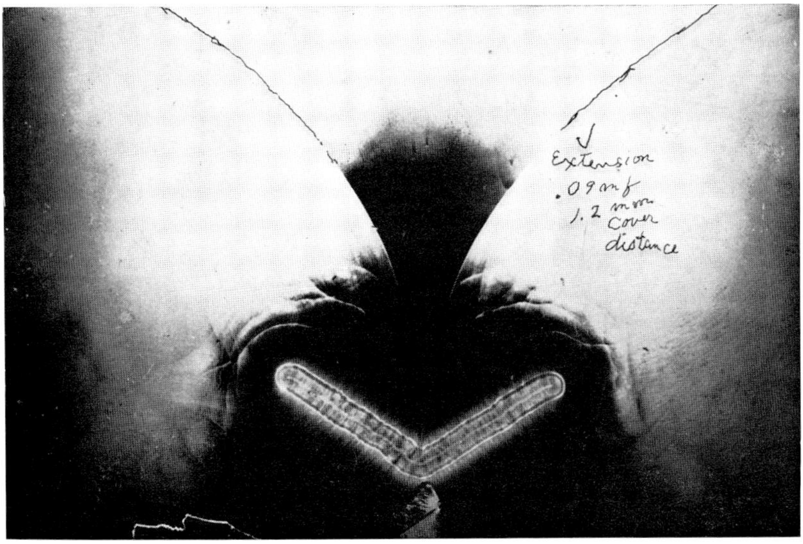

Fig. 5b. Soot funnel for Mach V-wide angle. (R. W. Wood)

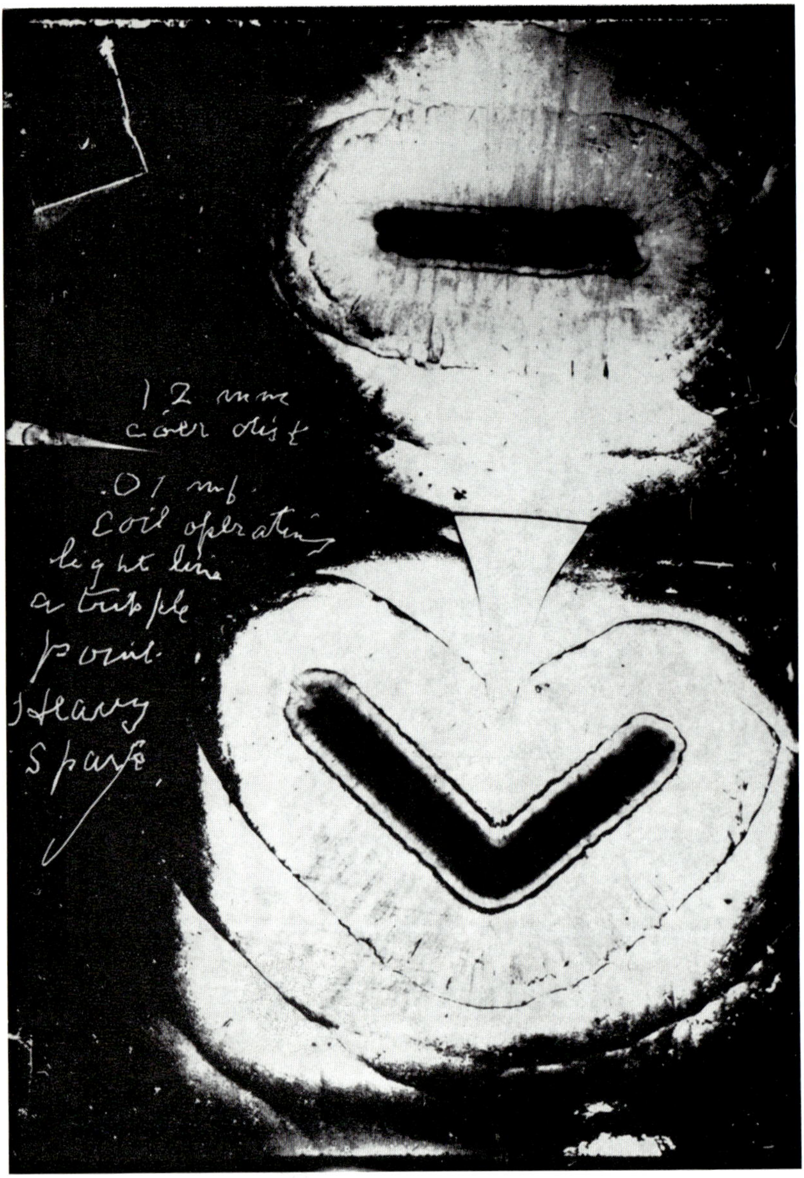

Fig. 6. 'Instantaneous' soot picture of two intersecting waves from a Mach V. (R. W. Wood)

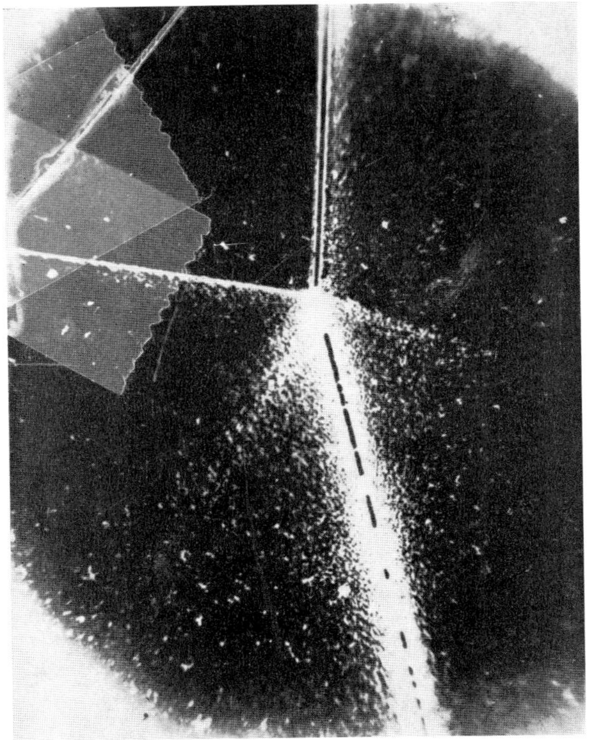

Fig. 7a. 'Instantaneous' soot picture (magnified) of point of intersection of Mach bridge and funnel. (R. W. Wood)

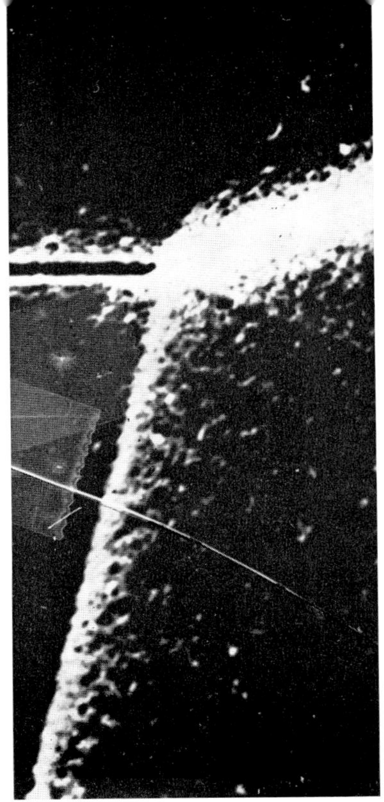

Fig. 7b. 'Instantaneous' soot picture (highly magnified) of point of intersection of Mach bridge and funnel. (R. W. Wood)

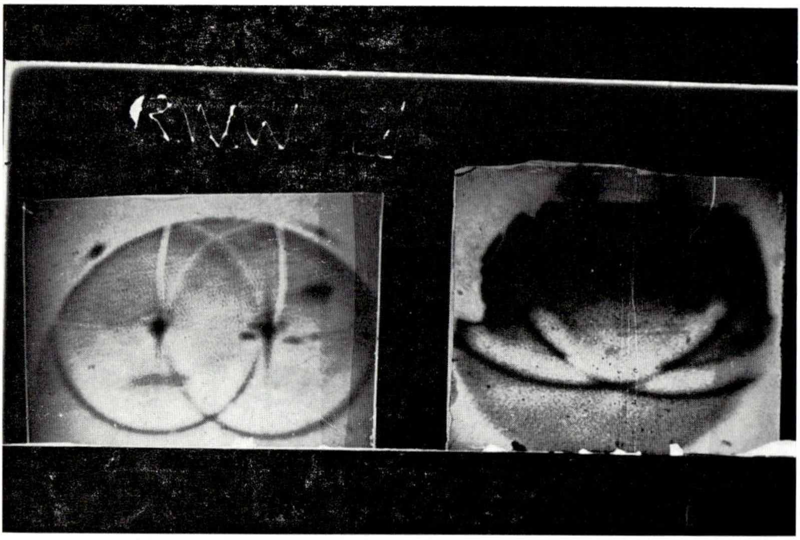

Fig. 8a. Schlieren photographs of two interacting (spark) shockwaves – early state. (R. W. Wood)

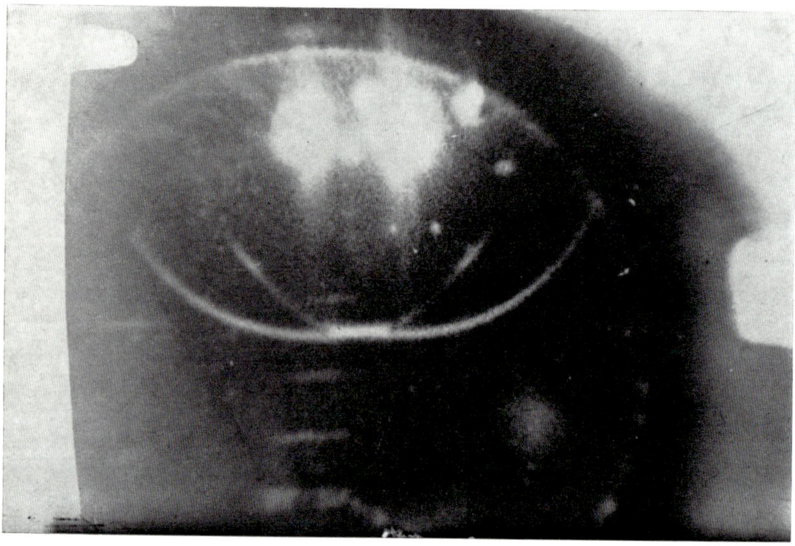

Fig. 8b. Schlieren photograph of two interacting (spark) shockwaves – middle stage. (R. W. Wood)

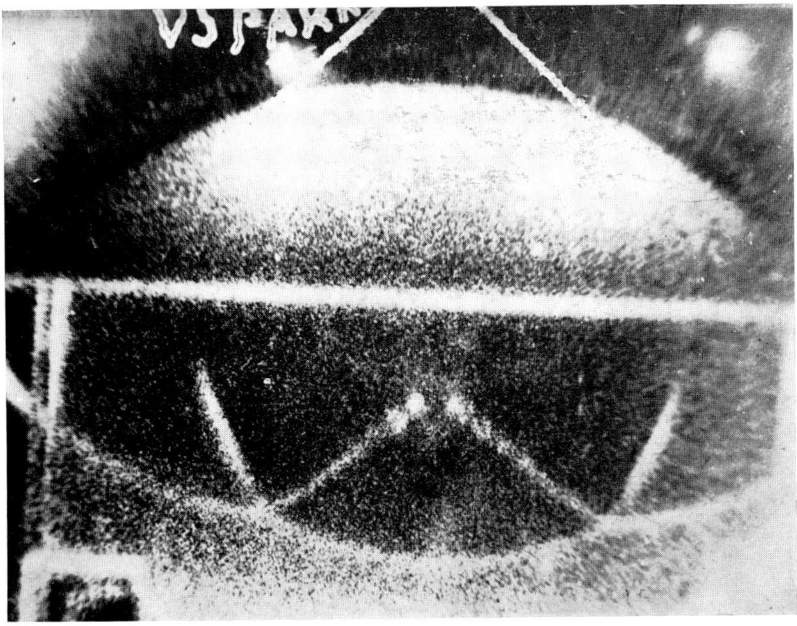

Fig. 8c. Schlieren photograph of two interacting (spark) shockwaves – late stage. (R. W. Wood)

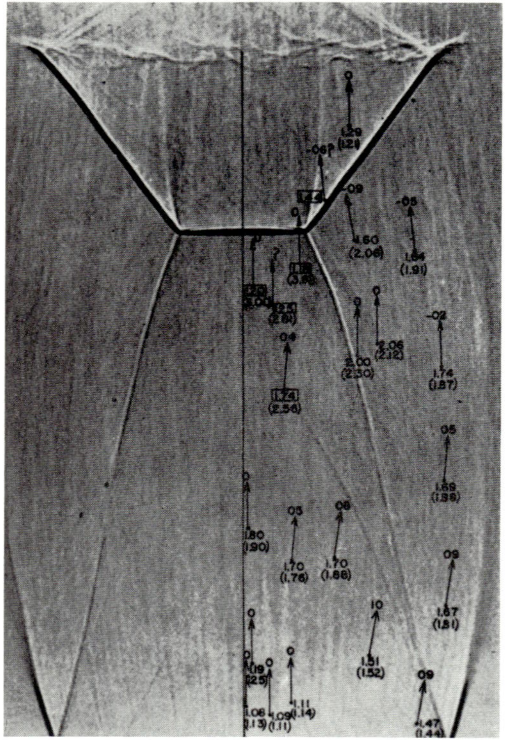

Fig. 9a. Shadowgraph of intersecting shockwaves in a supersonic air jet. (R. Ladenburg *et al.*)

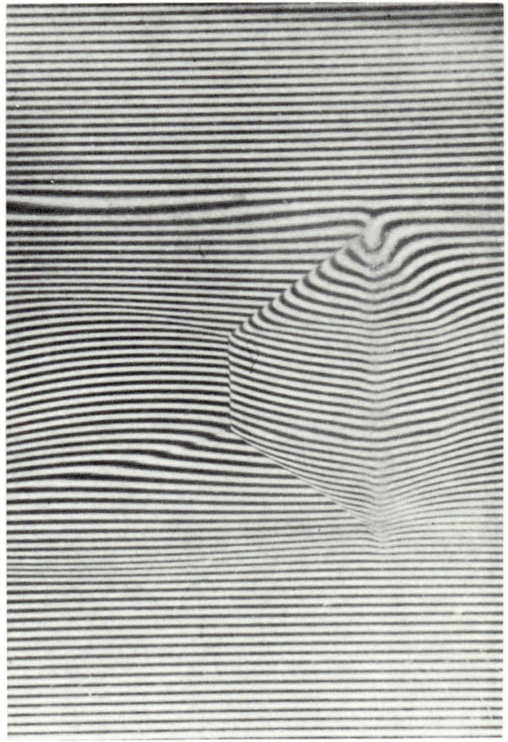

Fig. 9b. Interferogram of intersecting shockwaves in a supersonic air jet. (R. Ladenburg *et al.*)

vation, that is, to render palatable to the senses, something which we have only theoretically excogitated or theoretically surmised."[1e] And he remarked generally, "Frequently it is some unobtrusive idea which is productive of so much fruit and such extensive transformation in physical technics."[1f] Of even greater significance, however, was the necessarily continuing interaction of theory and experiment. In his Prague inaugural lecture, calling attention to the relationship between facts and ideas emphasized by William Whewell (1794–1866), he pithily concluded, "Ideas alone lead to empty speculations; mere facts can yield no organic knowledge."[1c] In the Vienna inaugural address, too, he argued, "Purely experimental inquiry, moreover, does not exist for as Gauss says, virtually we always experiment with our thoughts. And it is precisely that constant, corrective interchange or intimate union of experiment and deduction, as it was cultivated by Galileo in his *Dialogues*, and by Newton in his *Optics*, that is the foundation of the benign fruitfulness of modern scientific theory as contrasted with that of antiquity, where observation and reflection ofttimes pursued their respective courses like two strangers".[1a] Theory, moreover, was to be primarily a means, not solely an end in itself: "The beauty of the theory is, that it bears upon its face the stamp of truth ... The inquirer seeks the truth."[1h]

This youthful (1867) enthusiasm of Mach, however, was given up later; the absolute quest for eternal truth gave way to the relative guidance of pragmatic developments. As he admitted in 1871, "The object of natural science is the connexion of phenomena; but the theories are like dry leaves which fall away when they have long ceased to be the lungs of the tree of science."[1i] Later (1882) he limited "the goal which it [intellect] has set itself as the simplest and most economical expression of facts." Max Planck (1858–1947), it is true, was vehemently opposed to Mach's view inasmuch as he felt that "all great masters of research in science were moved by the fixed belief in the reality of their theory of the world picture and were not motivated and controlled by the principle of thought economy."[4] Evidently confusion exists here in the search for the understanding of phenomena and any belief in the finality of the current understanding. Critical exploration, with interaction of theory and experiment, results in theory being regarded more as a means than an end in itself. As Philipp G. Frank (1884–1966) remarked with respect to Mach's philosophy shortly after his death, "My view is that their [Mach's doctrines'] main value [for physicists] is not to help the physicist to go forward in his physical work, but rather that they provide the means for defending the edifice of physics against the attacks from the outside."[12a]

In conclusion, I wish to note certain unanswered questions, as well as some social implications, of Mach's work on shockwaves.

Frank concluded early that Mach's teachings were "stimulating and provocative".[12] In the field of shockwave phenomenon this prediction was certainly realized in World War II. The so-called Mach effect is still largely an unexplained phenomenon.[13] Even Ludwig Prandtl (1875–1953) himself concluded wrongly, "The theory of the forked shockwave has been put forward several times independently ... The phenomena which occur when explosion waves meet, discovered by E. Mach in 1875, and observed since, are related to the forked shockwave; an explanation of them has been recently given by F. Schultz-Grunow"[14] (1948). It is true that in the case of relatively strong shockwaves a simplified theory for three-shock intersections has been successfully applied by various individuals. Even to this day, however, there is no adequate general theory for the interaction of weak shockwaves.[15]

Mach began his lecture on the 'Photography of Projectiles' with the following humanistic remark, "To shoot, in the shortest time possible, as many holes as possible in one another's bodies ... seems to have risen to the dignity of a duty with modern men, who ... are bound by the equally holy obligation of making these holes as small as possible, and, when made, of stopping them up and of healing as speedily as possible."[1e] After discussing the physical aspects of the problem he then felt impelled to comment, "No one who has had the opportunity of studying modern guns and projectiles and their marvelous perfection, their power and precision, can help confessing that a high technical and scientific achievement has found its incarnation in these objects. We may surrender ourselves so completely to this impression as to forget for a moment the terrible purposes they serve. Permit ... a few words on this glaring contrast."[1e] He went on to conclude, "It requires a strong belief in the insufferableness of medieval barbarism to expect no great improvement."

Ironically, the Mach effect, it may be recalled, was actually used in the bombing of Hiroshima, that is, in determining the position of the atomic bomb best suited for optimum (?) damage. Hence we today must share in the pessimism of Mach's misgiving of long ago. At the same time, however, noting the international interest in his shockwave experiments, we should perhaps join also in his parting optimistic hope: "By the side of the problems which separate nations, the great and common ideals which claim the exclusive powers of the men of the future appear one after another in greater distinctness and in greater might." In this spirit we reconfirm Frank's expressed belief that "he [Mach] is still very much alive today."[12b]

U.S. National Science Foundation

REFERENCES

[1] Ernst Mach, *Popular Scientific Lectures*, transl. by T. J. McCormack, 5th ed. Open Court, La Salle, Ill., 1943.
[1a] 'On the Part Played by Accident in Invention and Discovery'.
[1b] 'The Fibres of Corti'.
[1c] 'On Transformation and Adaptation in Scientific Thought'.
[1d] 'On the Economical Nature of Physical Inquiry Projectiles'.
[1e] 'On Some Phenomena Attending the Flight of Projectiles'.
[1f] 'The Velocity of Light'.
[1g] 'Why Has Man Two Eyes'.
[1h] 'On the Causes of Harmony'.
[2] Ernst Mach, *Erkenntnis und Irrtum*. J. A. Barth, Leipzig, 1905.
[3] Ernst Mach, *The Analysis of Sensations*, transl. by C. M. Williams, 1st German ed., revised by S. Waterlow, 5th German ed., Dover, New York, 1959.
[4] Carlton Berenda Weinberg, Mach's Empirio-Pragmatism in Physical Science, Ph.D. Thesis. Albee, New York, 1937.
[5] Ernst Mach, *The Science of Mechanics*, transl. by T. J. McCormack, 6th ed., revised by K. Menger, 9th German ed. Open Court, La Salle, Ill., 1960.
[6] Ernst Mach, 'Über einige Wirkunge des elektrischen Funkens'. *Sitzungsberichte der Kaiserlichen Wien Akademie der Wissenschaften* **72**, 44, 1876.
[7] W. D. Kennedy, 'Explosions and Explosives in Air', in Summary Technical Report of the NDRC, Division 2, I. *Effects of Impact and Explosion*. Washington, D.C., 1946.
[8] John von Neumann, 'Oblique Reflection of Shocks', Washington, D.C., *U.S. Navy Bu. Ord. Explosives Research Report #12*, 1943.
[9] R. W. Wood, 'On the Interaction of Shock Waves', OSRD Report No. 1996, 1943. The photographs in this article are from the residue of Wood's scientific materials at the Rowland Physics Laboratory, courtesy of J. Wosyka Mandansky.
[10] R. Ladenburg, C. C. Van Voorhis, and J. Winckler. 'Interferometric Study of Supersonic Phenomena I, II', Washington, D.C. *Naval Ord. Reports* 69–46, 93–46; 1946.
[11] Ernst Mach, *History and Root of the Principle of the Conservation of Energy*, transl. P. E. B. Jourdain, Chicago, Ill.: Open Court, 1911.
[12] Philipp Frank, in *Between Physics and Philosophy*. Harvard University Press, Cambridge, 1941. (a) 'The Importance of Ernst Mach's Philosophy of Science for Our Times', 1917; and (b) 'Ernst Mach and the Unity of Science', 1938 [reprinted in this volume – Ed.].

[13] R. J. Seeger and A. B. Tayler, 'Triple Shockwave Interactions', *Phys. Letters* **2** (1962) 339.
[14] L. Prandtl, *Essentials of Fluid Dynamics*, transl. 1949, 3rd ed. Hafner, New York, 1952.
[15a] H. Polacheck and Raymond J. Seeger, 'Shockwave Interactions' in *Fundamentals of Gas Dynamics* (ed. by H. W. Emmons), Princeton University Press, Princeton, 1955.
[15b] D. C. Pack, 'Shockwave Phenomena' in *Research Frontiers in Fluid Dynamics* (ed. by R. J. Seeger and G. Temple), Wiley, New York, 1965.

PETER G. BERGMANN

ERNST MACH AND CONTEMPORARY PHYSICS*

Ernst Mach was a competent experimental scientist. He was also a great philosopher of science. As a young student in Prague I admired his schlieren photographs of rifle bullets in flight and of the shock waves accompanying them, photographs that were still on exhibit in the hallways of the physics institute whose director he had been in his time. These photographs had been taken in the light of spark discharges triggered by the bullets themselves. Professor Philipp Frank and Professor Rudolf Carnap taught us in Mach's spirit. Later on, at the Princeton Institute for Advanced Study, I was privileged to listen to Professor Einstein's reminiscences of his encounters with Mach, both through Mach's writings and through face-to-face meetings. Much as Einstein tended to disagree with Mach's philosophical views, he fully acknowledged the stimulus he had received as a young man from the unorthodox approaches to mechanics and gravitation pioneered by Mach.

In this talk I shall not dwell on Mach's contributions to our present knowledge of supersonic aerodynamics. This is not my field, and the subject is discussed by R. J. Seeger and W. F. Merzkirch. I should rather concentrate on Mach's style of critique of physical theories, and how critiques in his spirit may contribute to contemporary problems. I prefer this interpretation of my assigned topic to a detailed discussion of Mach's specific contributions to physical concepts, which include, for instance, the so-called Mach's Principle, or of his specific failings, such as his reluctance to accept the atomic hypothesis of matter and the theory of relativity. We cannot separate beyond doubt a man's personal idiosyncracies from his professed philosophical beliefs, and proclaim, *ex post facto*, that Mach's failure to grasp the significance of the atomic hypothesis was, or was not, the result of his peculiar epistemological and ontological views. I should like to discuss with you Mach's emphasis on empirical data, on observations, as the only secure foundations of scientific theorizing. To what extent has his attitude helped or hindered scientific conceptual developments, and what can we hope to gain from it today?

Boston Studies in the Philosophy of Science, VI.

Let me begin with an instance in the history of theoretical physics in which Mach's type of critique has proved eminently beneficial. I am referring to the development of modern quantum physics at the hands of E. Schrödinger and W. Heisenberg in 1925. In that year two apparently disparate theories of atomic systems were developed. Schrödinger's was based on the elaboration of De Broglie's hypothesis of matter waves. Heisenberg started out from the notion of non-commutative quantities in modern algebra. The following year, 1926, Schrödinger and C. Eckart showed, independently of each other, that the seemingly distinct theories differed only in terms of the intuitive pictures evoked by their respective authors, and that once reduced to their mathematical bones and to their statements about observations and measurements Schrödinger's and Heisenberg's quantum theories were equivalent.

Mach's injunction, to analyze a physical theory not in terms of its 'metaphysical' superstructure but in terms of its relationships to observations, emerged triumphant. In fact, the accomplishments of the year 1926 have tended, I believe, to influence to a very large extent the subsequent epistemological history of quantum physics. We have been enjoined to analyze quantum-physical statements not so much in terms of what they tell us about the objects of our interest, i.e., the physical systems before us, but rather in terms of what they tell us about our observations and measurements, about 'sense data'.

Much of the epistemological discussion in our century has centered about the dispute between 'positivists' and 'realists' or 'materialists'. Crudely speaking, the realist is convinced of the existence of the physical universe about us, and he considers our experiments, observations, and measurements as the means to discover the properties of these surroundings. Positivism, in its pure form, denies that inquiry into the reality of the universe, its existence independent of our observations, possesses any legitimate meaning. It decries such inquiry as 'metaphysical' and insists that the objective of our science is to systematize our experience, to discover persistent traits and regularities and, perhaps, to predict, to anticipate, the results of experiments not as yet performed. Clearly, Mach must be grouped with the positivists.

For a long time it might have been argued that the distinction between the primacy of objective reality, as claimed by one school, and the primacy of empirical information, as claimed by the other, represented more an emotional than a logical difference, and that the realists and the empiricists would pretty much do the same things in the scientific laboratory or in the course of interpreting recorded data. The very fact that the young Einstein could be deeply impressed by Mach's critique of classical mechanics, and yet reject his philosophy almost completely in his later years indicates that Mach's contributions were in many respects valid beyond the framework of his philosophical views. One could, perhaps, argue that Mach analyzed physical theories so as to discover and to separate their physical substance from their doctrinaire embellishments. If by embellishments we are to understand conceptual elements which have no bearing on ascertainable facts, then both the positivists and the realists can accept and welcome such an analysis.

Developments of this century have added a new dimension to the ancient philosophical discussion. I am referring to the fact that man's experimentation necessarily modifies the object which he is exploring. The fact that observation also implies interaction has become significant of recent years. First, man's technological power has grown to the extent that increasingly measures intended only to increase comfort or to obtain information will modify our environment far beyond the original purpose. Examples come readily to mind. Testing nuclear bombs increases ambient radioactivity so that it becomes a hazard for life on this planet. Burning fossil fuels at the present rate modifies the composition of our atmosphere so much that long-range climatic changes may result. Exploding even very few nuclear bombs in space changes the composition of particles in near-space so that for years the radiation belts surrounding the earth are profoundly altered. Our industrial wastes threaten to choke life in the fresh waters of all the continents on which our Western civilization has taken hold; and this is merely one instance of pervasive ecological change resulting from human activities.

Clearly, as man's tools become ever more powerful, not only industrial and other mass activities change our environment. We must now consider in increasing measure the possible long-range effects that some experiments, performed for the sake of scientific research, might have on our whole environment. The social and physical effects of scientific experimentation assume a new dimension; and this new dimension has not only ethical but also epistemological implications. The very process of inquiry now modifies the object that is being inquired into. In order to retain any perspective on the nature of scientific exploration we must differentiate between the object as it would exist without our experimentation and the object modified by our experiment. The psychologists have long known that individuals respond to explorative procedures by changing. Physical scientists must learn that their objects of inquiry may do the same; and that the changes may be sufficiently universal that by the very process of experimentation we may sometimes destroy that which we wish to know better. All that I have just stated implies, however, that the objects to be investigated exist and possess properties, prior to our exploration, properties that may or may not be affected by our experiments.

Observational procedures will modify the object of our interest also at the atomic and the subatomic level. And at that level even the most circumspect choice of experimental procedures cannot avoid that the object is changed by the interaction that we must set in motion in order to obtain information. It lies in the nature of quanta that the means by which information about a physical object might be obtained cannot be thinned out below a certain strength; hence the object will scatter a quantum of a certain kind, and it will suffer a recoil, so that after the performance of a measurement it will have changed some of its characteristics in an unknown and uncontrollable manner. Consequently, there is no experimental arrangement by which we can gain knowledge simultaneously of all the characteristics of an atomic system which separately can be ascertained without difficulty. This state of affairs is generally known as Heisenberg's uncertainty principle.

Philosophical realists have posed the question whether an electron, for instance, 'possesses' both location and linear momentum at the same time, so that we are merely unable to determine the numerical values of both of these variables simultaneously, or whether the electron by itself cannot be said with assurance to have any of these properties unless it has been 'prepared', for instance by the performance of a measurement, to be in a state endowed with an appropriate selection of properties. From the positivist's point of view this is a meaningless question. The theory of quantum systems enables us to predict the outcome of experiments, given the knowledge of certain antecedents; this is all that may be required of a physical theory. Certainly, at the atomic level the predictions of quantum theory have been verified by several decades of assiduous experimental research to any degree of accuracy and reliability that can be asked for.

But there remain a number of nagging questions. I have tried to juxtapose the problems raised by experiments on a very large scale and those raised by experiments on the microscopic scale. In both cases the process of experimentation tends to alter the object to be observed. But there is a difference. On the large scale we can hope to design experimental techniques in which this effect is either fully taken into account or, better yet, reduced to insignificant proportions. On the atomic level such a possibility is excluded as a matter of principle, or at least according to our present state of understanding. Hence the quest for 'reality' of certain properties of the physical substrate takes on somewhat different aspects on the macroscopic and on the microscopic scales. These are, however, not the only points of distinction.

I should like to direct your attention to two points. One is the need to perform a critique of what is meant by 'observational data'. The other is the statistical nature of the statements of quantum theory. Both, I believe, are matters that are suggested by the kind of analysis that Mach applied to the physics of his time.

In everyday life an inquiry into what we mean by 'observation' appears frivolous, or at least hair-splitting. And in many areas of natural science too, it appears perfectly clear to all concerned what is meant by observation. The astronomer observes celestial phenomena by means of a telescope, and he interprets and refines his observations in terms of his knowledge of the optical properties of his apparatus as well as of the atmosphere. But as we approach the microscopic level, the link between the raw data and our conclusions becomes more extended and less direct. In a conceptual experiment we may talk about 'locating' the electron within an atom by means of an X-ray microscope. In the real laboratory experiment we do no such thing. We perform complex scattering experiments not on one but on a large number of like atoms, and we interpret markings on photographic plates, or counts by an electronic system, in order to gain information on the location of the electron in question. Just how we utilize the readings of certain macroscopic instruments to arrive at interpretations, saying that we have 'observed' a given property of an atomic system, is an integral part of our theoretical scheme of thing.s

Let me give you another example, which in some sense gives us a direct tie with Ernst Mach's writings. Mach was dissatisfied with the classical definition of inertial systems, because it appeared to be based on absolute, and 'non-observable' properties of space and time as such. He proposed that the inertial properties of terrestrial test bodies were the result of interaction with the very large masses at great distances, that is to say the distant stars and galaxies. It did not occur to Mach that the motions of the very large and very distant masses are known to us only by way of electromagnetic waves reaching us and that the interpretation of these light signals involves hypotheses concerning the Euclidean nature of space and the straight-line propagation of light rays through that space. Of course, hindsight is cheap. Today we are aware of the substantial nature of these hypotheses because we have been educated by the theory of relativity to be alert to alternative possibilities in this regard.

What appears to me significant is that to single out the motion of the distant masses as something observable is an integral part of the structure of a physical theory that we might identify as classical celestial mechanics. Other theories involve different assumptions as to what is 'observable', and usually these assumptions are non-trivial.

Permit me to stay with 'Mach's Principle' a little longer. Through the development of electromagnetic theory and of general relativity we have been accustomed to the idea that not only material particles are the stuff of which physics is made but also fields. Maxwell was not the last physicist to retain a suspicion that these fields might be artifacts, a convenient means for explaining the motions of particles under conditions in which the tenets of Newtonian mechanics were no longer valid. For the electromagnetic field, Fokker and Tetrode, and in more recent times Wheeler and Feynman, have erected structures in which the electromagnetic field appears merely as an auxiliary device for the treatment of particle motions. Occasionally one can read that whereas particles can be observed 'directly', fields can be probed only by way of the ponderomotive effects that they have on particles, in other words, that particles can be observed but not fields. Is this true?

Very obviously, we do not observe particles 'directly', either. If we do observe them visually, we do so by way of the electromagnetic fields entrained or scattered by the particles. If we observe particles through photographic or counting devices, we utilize even longer chains of cause-and-effect relationships to interpret the indications of our instruments as referring to particles and their modes of behavior. Should we state that we observe fields more directly than particles? I do not believe that such a statement is justified any more than its reverse. True, electromagnetic waves enter our eyes. In so doing they interact first with the material particles constituting the eye's refractory system (or no image would be projected on the retina), then with the intricate physico-chemical system of the retina. The resulting excitation is propagated along the optical nerve by a mechanism that surely is not an electromagnetic wave. In other words, as information is transmitted from the material particle to our central nervous system, a huge variety of intricate linkages is pressed into service.

Einstein has emphasized that any act of cognition involves concepts previously formed. Concept formation is logically an act of free creation, but subject to validation in the light of experience. An infinite chain of concepts stretches back to the earliest experiences of the newborn baby, which can never be disentangled and fully analyzed. The workings of our mind and the physical means of communicating between the individual and his surroundings combine to make the seemingly simple act of observation one of utmost physical and conceptual complexity. This complexity does not invalidate observation as the mainstay of science. Only, as Einstein has pointed out time and again, observation is no more a monolithic 'indefinable' than any other part of what we call scientific exploration. It requires analysis and continuing critique. And back of the observation is the something that is the object of observation, the universe.

Because we cannot always deal with the full complexity of nature, because we cannot ask all the questions worth asking at once, we generally hypothesize in any physical theory that certain basic things are observable. These assumptions, as I have said, are an integral part of the theory, and like the rest subject to review and modification. As long as we remain aware of the ephemeral character of the specific assumptions of observability made, I see nothing wrong with them; in fact I believe that such assumptions are unavoidable in constructing any physical theories at all. The question is only whether we are aware of what we are doing.

Returning once more to Mach's principle, I should say that in Mach's time it was sensible to assume that the locations and motions of the distant galaxies in the flat three-plus-one-dimensional universe were observable and that the general laws of mechanics should be expressed in terms of interactions between all the masses. If this program turned out to be successful, then the elimination of the notion of *a priori* inertial systems was a step forward. In the meantime, general relativity has suggested that the gravitational metric field is capable of steering the material bodies immersed in it, but that the converse does not hold. The vacuum field possesses degrees of freedom not tied to the motion of any massive sources. Hence I fail to see the attractiveness of Mach's principle from the point of view of today's understanding of gravitation and of ponderomotive laws in general.

This is not to say that in field theory there exists no analog to Newton's inertial systems. Analogs are provided by the imposition of boundary conditions on the field in addition to the local laws, the partial differential equations known as the field equations. But the analogy fails to be perfect in several respects. Though one can impose boundary conditions that lead to the uniqueness of inertial systems, there are also known possible alternative boundary conditions that are considerably less stringent. And because of the considerable and interesting variety of boundary conditions that one can conceive of, the selection of a particular set becomes a non-trivial objective of scientific inquiry. Relativistic cosmology contains the investigation of possible boundary conditions as an integral part. And what makes the motions of the distant galaxies most remarkable is that there is a pattern to these motions, a pattern that we are still far from understanding.

I shall now turn to the aspect of microphysics that distinguishes it from the corresponding situation in macrophysics, the statistical nature of its pronouncements. According to quantum theory, the maximal information that can be validly gathered about an atomic system determines its 'quantum state'. Given this maximal information, one can predict the outcome of some measurements that might be performed on the system, but these represent an insignificant fraction of all the observations that one might contemplate. As for the remainder, quantum theory predicts for every observation the set of all possible results, and further for each of these possible results the probability of its actual occurrence.

Probabilistic predictions have, of course, no meaning for the individual system. If one individual system is subjected to a specified measurement, there can be but one outcome; the rub is that the theory does not tell us with certainty which one it is going to be. The probabilities of quantum theory have this significance: If we prepare, or select, a large number of systems of the same kind so that all of them are in the same specified quantum state, then the performance of the same measurement on each system will yield diverse results, and each numerical result will occur in a percentage of cases given by numerical probabilities characteristic of quantum predictions.

There are many situations in atomic physics in which each experiment involves a large number of like systems to begin with. Quantum-theoretical probabilities may then be translated directly into intensities of spectral lines, intensities of split beams of particles, etc. It is in cases like these that quantum theory proves its power of predicting both the location and intensity of spectral lines, or the shape of the absorption spectrum of a crystal, and similar experimental results. These are, however, not the only situations ever encountered in atomic research. When it comes to research in the area of elementary particles, the experimenter very frequently observes individual occurrences as such, and he laboriously, and at great expense to the taxpayer, accumulates them into the large sets of like systems of which the theory deals.

Quantum theory then presents us with a very strange situation. It tells us something about 'observations', but it is concerned not so much with objects and with observations on objects, but with sets of objects and with observations on sets. In classical physics a law of nature was thought to describe a completely repetitive situation. Given the antecedents, the 'cause', the resulting 'effects' would invariably be the same. Quantum laws equally assert repetitiveness and reproducibility. But what is reproducible is not the individual case history but the dispersion of large sets of case histories.

Very obviously there is involved here a whole set of assumptions as to what are proper sets of case histories. With the benefit of hindsight one can accumulate case histories which show almost any dispersion desired. If I search diligently among the life histories of the population of the United States, I can discover a thousand people who were born in 1900 and who died exactly thirty years later; and I can discover another thousand individuals who were born in 1910 and lived exactly forty years. It would be premature to conclude that the lifespan of people increases ten years every ten years. For purposes of statistical research one must select samples 'at random'. Statistical statements of quantum theory refer to samples which are randomly selected in a certain sense, which I believe we do not yet understand completely.

At any rate, though statistics also plays an important role in macrophysics, its unavoidability is peculiar to the atomic domain. It is necessary to carry out experiments on large numbers of equivalent systems, not in order to be able to average out random errors but because the quantum laws are concerned with populations of systems. Somehow the built-in dispersion is related to the interference that every observation represents. One can argue that the outcomes of experiments are diverse because their performance requires the exposure of the objects of our interest to the interaction with other physical objects, such as light quanta, under circumstances in which the initial relationships between the interacting objects are not precisely reproducible. This may be quite true; but the fact remains that this diversity, whatever its root cause, cannot be eliminated by any stratagem known. The performance of an observation generally changes the object of our observation in an uncontrollable manner; and thus in quantum physics the distinction between action and contemplation is blurred beyond repair.

Perhaps the present state of quantum physics simply reflects a fact of life even in our interactions with inanimate nature. What the scientist of today can hope to learn from Ernst Mach, I believe, is not so much Mach's conservative resistance to the then new theories of atomism and relativity, but his plea that the scientist should not fall victim to petrified notions, that he should be willing to apply his critical faculties to every aspect of the process of scientific inquiry.

Syracuse University

REFERENCE

* This work has been supported by the U.S. Air Force Office of Scientific Research.

ERWIN N. HIEBERT

THE GENESIS OF MACH'S EARLY VIEWS ON ATOMISM*

It is appropriate to consider three distinct phases of Mach's changing views regarding the ultimate structure of matter. As a young 'Dozent' at the University of Vienna, until 1863, Mach accepted the atomic theory without reservation – at least as a useful working model and hypothesis. His earliest views on atomism are contained primarily in the investigations on the Doppler effect, capillary phenomena, spectra of gases, resonance vibrations, and the transmission of sound in the ear. In addition Mach employed the atomic theory in his compendium of physics for medical students which was published in 1863.

By 1864, when he moved to Graz to accept the professorship in mathematics there, Mach's views on atomism obviously were under serious revision. The first explicit and sharp criticism of atomic conceptions was given in his little book of 1872 on the history and root of the principle of conservation of energy. Strong anti-atomistic comments also appeared in his major treatises in the 1880's and early 1890's. During the last two decades of his life Mach made some attempt to reformulate his views. Reluctantly he made an attempt to reconcile his opinions with the new evidence which was being furnished in support of the kinetic-molecular theory of matter. In any case, toward the end of his long and fruitful career as scientist, historian, and philosopher, Mach became much less outspoken in his denunciation of atomic conceptions. Although he conceded on a number of occasions that the theory had demonstrated its usefulness, it is doubtful whether Mach ever at heart really gave up his anti-atomistic presuppositions.

The object of this paper is to analyze those early works of Mach in which we find support for the atomic theory in one way or another.[1] The account begins with some incidental remarks in conjunction with the then current disputes concerning the Doppler effect. Mach wrote three papers on this subject between the ages of 22 and 24. Ten years later when these papers were reprinted together in pamphlet form he remarked that his earlier views on the subject were fundamentally sound in spite of the fact that they revealed all the typical shortcomings of a self-taught beginner.[2]

Briefly, the background to Mach's papers on the Doppler effect was as follows. In 1843 Christian Doppler of Prague had published a paper in which he maintained that the color of light as well as musical pitch are altered in consequence of the rapid motion of the sources of light and sound relative to the observer.[3] The Dutch meteorologist Buijs Ballot conducted some elaborate railway experiments and verified the Doppler effect with respect to sound in 1845.[4] In 1852 Joseph Petzval, Professor of mathematics at the University of Vienna, on theoretical grounds attacked Doppler's views from the standpoint of his own "law of the conservation of the period of oscillation".[5] Ångström of the astronomical observatory in Uppsala claimed in 1853 that his experiments on metallic arc spectra did not support Doppler's conclusions for light.[6] Doppler died in 1853 while the controversy was in full swing.

In 1860, a few months after he had completed his doctoral degree, Mach presented an experimental and theoretical defense of Doppler's law against the criticisms of Petzval and Ångström.[7] Included in this work was the description of an ingenious laboratory device to demonstrate the change of frequency of sound as a function of the speed of motion of the source. The investigations were carried out in the physical institute of his teacher, Andreas von Ettingshausen, who had taken over Doppler's position as Professor of Physics at the University of Vienna in 1853. While it was not feasible at this time to carry out a similar laboratory demonstration for light sources[8] Mach saw no theoretical reason to assume that the effect should be restricted to sound waves. He therefore also supported Doppler's conjecture with respect to light.

In the course of discussing the effect Mach accused Petzval of resting his objections to Doppler's law on analogical reasoning in place of *bona fide* investigations. He maintained that Petzval's arguments were vitiated by the failure to distinguish between the motion of the waves and of the particles (*Theilchen*) composing the medium which is set in vibration by the source. Similarly Mach argued that Ångström's rejection of Doppler's ideas was based upon an erroneous interpretation of his experiments showing the absence of any spectral shift for the electric sparks from metallic electrodes. "Ångström", said Mach, "confuses the advance of the glow with the advance of the glowing particles, and that is just as inadmissable as to confound the advance of a water wave with the advance of the water particles."[9]

In another paper on this subject, in 1861, Mach tried to show that Petzval's theoretical deductions from the principle of the conservation of the period of oscillation had nothing to do with Doppler's experimental results.[10] In comparing Doppler's law and Petzval's principle Mach wrote: "Both principles relate to very different cases and the one can no more be substituted for the other than an armchair for a taxi."[11]

Although Mach had nothing to say directly about atoms in these papers, the assumption of vibrating particles of air (for sound) and particles of ether (for light) lies at the root of his criticisms of Petzval's and Ångström's interpretations. And, as in the case of his explicit atomistic interpretation of chemical spectra (discussed below) so here, Mach accepted the Doppler shift as a real displacement of spectral lines. The displacement was seen to be the effect of motion on the *physical* conditions of the vibrating particles of the wave-transmitting media. Mach was the first to explicitly suggest the spectroscopic determination of the motion of luminous bodies based on the Doppler effect.

In a paper delivered in 1862 Mach discussed the merits of various theories (e.g., as contained in the works of Young, Laplace, Poisson, and Gauss) for deriving the so-called 'molecular function' of fluids in relation to capillary phenomena and the principal radii of curvature of liquid surfaces.[12] In conformity with common opinion at the time, Mach believed that the surface shape of a liquid was determined solely by the intermolecular forces between the particles. He suggested, however, that perhaps the known discrepancies between experiment and the simple 'molecular force' theory might be met by postulating repulsive forces at very small distances in addition to attractive forces falling off rapidly as a function of intermolecular distance. In any case, Mach proceeded to reason on the assumption that the molecules exerted forces on one another according to some as yet undetermined function of their distances of separation.

His investigations, accordingly, were designed with the object of studying the molecular forces of liquids under experimental conditions as far removed as possible from the simple equilibrium equation, $1/\varrho + 1/\varrho' =$ Const., which relates the principal radii of curvature ϱ and ϱ' of a liquid at any point in the surface under conditions of the absence of the disturbing effect of gravity. Surfaces for which this relation holds had been shown by Gauss to be surfaces of minimum area. Mach attempted to ascertain, by optical measurement of the displacement of thin plain parallel fluid lamella in relation to their density, the law according to which "the individual molecules act upon one another" under various conditions. He recognized however that his experimental measurements fell far outside the limits of observational accuracy.[13]

The details of Mach's other experiments (none of them convincing, he thought) need not concern us here except to mention that he was manifestly preoccupied with a model of interacting molecules the intensity of whose forces were reducible to some function of the distances between the particles. In addition, he felt that it might be possible to relate the density and viscosity of liquids to their "molecular forces".[14]

It is noteworthy that Mach had begun his 1862 paper as follows: "The present note is to be regarded as the beginning of a long series of investigations concerning the molecular action of fluids – which investigations may be at least as important as the study of gases and crystals." The paper ended with the statement that he intended to study the various figures of the cohesion of fluids under a great variety of experimental conditions. But Mach never again came back to this subject from the point of view of molecular attractions.

When he took up the subject of "The Form of Fluids"[15] once more in a popular lecture in Prague in 1868, Mach was single-mindedly wrapped up in characterizing the external phenomenological features of the subject from the viewpoint of his principle of the economy of thought. Thus he illustrated the principle of minimum surfaces and the attending conditions of equilibrium by saying: "Nature acts in the construction of liquid figures on the principle of a covetous tailor (*eines habsichtigen Schneiders*), and gives no thought in her work to the fashions."[16] Or again:

> You see our miserly mercantile principle (*geizig kaufmännisches Princip*) is replete with consequences. The result of sober research, it has become as fruitful for physics as the dry questions of Socrates for science generally. If the principle seems to lack in ideality, the more ideal are the fruits which it bears.
>
> But why, tell me, should science be ashamed of such a principle? Is science itself [regarded as a maximum or minimum problem] anything more than – a business? Is not its task to acquire with the least work, in the least possible time, with the least possible thought, the greatest possible part of eternal truth?[17]

In a short note of 1862 Mach explored the possibility of a purely mechanical interpretation of the bright line spectra of chemical substances with the intention of eventually being able to use this approach "to understand clearly many points about the atomic theory".[18] Accepting "the well-established theory of gases of Krönig and Clausius", *viz.*, that there is no mutual interaction between molecules in the gaseous state, Mach reasoned that the relatively limited number of bright lines exhibited by gaseous emission spectra were to be correlated with the standing modes of vibration of the atoms in the stable molecular grouping. The spectral lines, that is, represented only the intramolecular vibrations. "For small amplitudes of vibration, i.e., for small excitations", said Mach, "the force of displacement acting in the atom can be set proportional to the displacement from the equilibrium position, so that the atom vibrates according to the simple law $x = a \sin kt$." While each atom was seen to possess only one period of vibration and to yield a single bright line corresponding to its wavelength, Mach indicated that the molecular grouping, as a whole and under varied external conditions, might offer periods of vibration for other possible positions of the atoms in the molecule. "At greater amplitudes, for more intense light, the forces no longer are proportional to the displacement and the vibrations become more complex. As experience shows, the spectrum becomes richer in bright lines." Mach conjectured that a law of atomic interaction might follow from an analysis of the spectral changes arising from the increased intensity of excitation – thus to provide information about molecular structure.

Underlying these considerations was a kind of Proutian hypothesis for the ultimate structure of matter. Mach said:

Gravity acts on all bodies alike. Thereby, in some measure it becomes probable that there is only one kind of matter-atom (*Körperatom*). Proceeding from this hypothesis, which certainly is worth testing, the atomic weight is fixed by the number of atoms which compose a gas molecule. The spectrum provides some information about the grouping; similarly the crystal form. The investigation of the relation between atomic weight, spectrum, and crystalline forms for chemically simple bodies appears to promise much.

From what has been said, chemically dissimilar gases would be distinguished from one another only according to the number and grouping of the atoms of the body … .

At least, in the further pursuit of these ideas there dawns the hope of seeing chemistry transformed into applied mechanics.[19]

In 1863 Mach published a predominantly experimental paper in which he sought to employ the theoretical principles of acoustics and resonance

– the Fourier analysis of tones and their overtones – to study the physiology of variations in blood pressure.[20] This work was an outgrow of experiments with the sphygmograph, an instrument for recording the rate, force, and wavelike variations of the pulse.

Of interest to us here is Mach's preoccupation with the interpretation of all kinds of wave motion in terms of mechanically vibrating systems. He wrote:

> One and the same problem in mechanics can give rise to very different questions. The problem of resonance ... can be applied directly to the phenomena of sympathetic vibration ... or to the variation of blood pressure as expressed in a given sphygmograph curve. ... If one treats the laws of resonance within this framework (but generalizes for a vibrating system) ... one can apply them [the laws] to the light of incandescent gases to determine, as I have shown [elsewhere], the construction of gas molecules and the state of the vibrating system.[21]

In another important and related paper of 1863, on the auditory organ, Mach sought to emphasize the close relationship between certain technical and physiological problems by treating the ear as a kymograph, i.e., as an instrument for recording its wavelike muscular contractions.[22] Implicit in the approach of this paper was Mach's attempt to explore the tones and their overtones by analogy to one-dimensional space; thus to demonstrate the feasibility of treating physical problems from the standpoint of functional relationships which are not bound by 'arbitrary' choices such as those of three-dimensional physical space. Years later, Mach was to explore the same means of 'describing' the properties of atoms and molecules in terms of any number of thinkable (useful) dimensions. Such a description, he hoped, would render the substantival and spatial characteristics of matter less cogent.

In this philosophically enlightening and fairly sophisticated mathematical analysis of the physiology of the ear as a kymograph, Mach formulated two separate problems in connection with the ear's function as a device for transmitting sound waves: *viz.*, (1) the analysis of the movements of the ossicles, or bonelike structures in the tympanic cavity; and (2) the problem of molecule-to-molecule impact propagated without any movement of the ossicles. Thus: "An easily moved rod, struck at its end, moves as a whole. If it is moved only with difficulty, then by contrast, the concussion is propagated by its molecules." The enormous differences in the moduli of elasticity were taken to be determinative for the two types of propagation of sound. Mach was unable to deal with these specific problems at this time except for a short mathematical note giving the displacement, from equilibrium and as a function of time, of two interacting molecules of given moduli of elasticity.[23] He suggested that he would return to an examination of his theory at such time as the experimental information for the moduli of the ossicles and their links became available; but he never did.

The remainder of the paper was devoted to exploring the mechanics of the ear in relation to action and reaction, resistance, surface area, lever motion, and conservation of energy. Were it not for the fact, mused Mach, that all of these physical measurements were undertaken on the strength of preconceptions regarding the ear's mode of operation, it might be possible at this stage to use the acquired information to fabricate an ear similar to the real human ear. Already skeptical about any attempt to reduce nature to purely mechanical motion – of ossicles or molecules – the young Mach added:

> But nature has not studied at the école polytechnique. Nature also has considerations to comply with other than just to ask permission from the prevailing theories of the times. Therefore it is doubtful whether she will make use of the suggestions of Savart, Seebeck, or my humble self. Perhaps nature is not able to employ the [above named physical] means in an unlimited way. Very likely she must stay within certain limits ... for other (anatomical) reasons.[24]

Groping for some alternate approach to a theory of motion of the auditory organs, Mach succeeded in using d'Alembert's principle to derive a fourth-degree equation with negative exponents – "a complicated expression of no interest, so long as experiment does not lead to special values of the coefficients". Manometric experiments designed to register changes in the ear drum resulting from various fixed tones also "failed altogether".[25] Mach concluded by philosophizing that a purely general theory of the auditory organs mostly yields only general results. "More the theory does not say, other than refusing to exhibit specific values for empirical constants. The theory therefore guides consistently toward experiment." Thus, "the ear must be assayed piece by piece. Herewith it will be indispensable to carry out experiments on anatomical preparations using arbitrarily constructed physical instruments".[26]

Mach published this paper at a time, it seems to me, when he was already critically examining the role of mechanism and atomism in the theoretical structure of science. Within a year Mach was devoting most of his attention to problems which could be attacked from the side of psychophysics – simultaneously pushing aside the empty mechanistic and atomistic interpretations which, so it seemed to him, had netted him next to nothing in his physical investigations even though they apparently had been of some value as an aid to teaching. Besides, at this time, Mach simply had not the means to acquire the instruments which were necessary for investigations into the more conventional problems of physics. Almost fifty years later he wrote:

On completion of my university studies, luckily or unhappily, I lacked the means for physical investigations and thereby was pressed into the field of the psychology of sensation. Here, where I could observe my sensations, and against their environmental circumstances, I attained, as I believe, a natural *Weltanschauung* freed from speculative, metaphysical ingredients.[27]

Notwithstanding these disappointments in his own experimental research, Mach by this time had made good practical use of the atomic theory in a course, entitled 'Physik für Mediciner', held three times per week at the University of Vienna during the winter semester of 1861/62. The lectures provided the basis for a manual of physics for medical students which was published after Mach had achieved the *venia legendi* and become *Privatdozent*.[28] The volume was dedicated to his teacher, Andreas von Ettingshausen, who had provided him with so many experimental opportunities and facilities in the physical institute and laboratory at the university. Von Ettingshausen's *Anfangsgründe der Physik* (1844), which supported the atomic conception of matter, was a much used university textbook and had gone through four editions by 1860. We shall see that Mach's use of the atomic theory in explaining the physical and chemical properties of matter in his textbook did not differ radically from the general expositions of his contemporaries. That is, the same general arguments were being invoked by scientists everywhere – whether they believed in the real existence of atoms and molecules or not.

In the preface to his *Compendium* Mach wrote:

The present system [of physics] is becoming more inconvenient by the day. It is no longer possible to represent light phenomena or heat phenomena *per se*; and this inconvenience must become more perceptible in future. The various kinds of molecular processes will then be treated systematically in one textbook of physics so that it will be merely incidental to consider them as light, heat, and sound which are seen, felt, and heard.

I have allowed the atomic theory to take first place everywhere [in this work], indeed, not in the belief that it represents the last word, or no longer requires additional support, but because it brings the phenomena into a simple and visual relationship. One may accept the atomic theory, if it is permissible to express oneself in this way, as a formula which already has led to many results and will continue to do so in future. In point of fact, whatever kind of metaphysical conception [theory] of matter may be put forward in future, the results acquired according to the atomic theory will always be capable of being translated into the other – just as formulas in polar coordinates can be expressed in parallel coordinates.[29]

In the purpose and plan of Mach's *Compendium*, the atomic theory therefore took its place as a kind of "philosophical accounting of physics". Thus he wrote: "After we have seen in our considerations of the atomic theory, that probably all physical phenomena are stipulated by the equilibrium and motion of the atoms or molecules, we may proceed to the general laws of equilibrium and motion, i.e., to mechanics."[30]

The obvious pedagogical value of the atomic theory, even in a textbook intended for medical students, did not by any means obscure Mach's epistemological position concerning the status of hypothetical constructs and their function in the reciprocal interplay between inductive and deductive processes in science. Presumably Mach was in no position at that time to offer a more credible or convenient alternative to the atomic theory. Nevertheless it is evident that he wanted his students and readers to recognize that his references to the atomic theory of matter were offered on the strength of a mere hypothesis.

Conceivably, the atomic theory, someday, as the result of new scientific experiments and new ways of looking at nature, would lose its place and relevance within science; i.e., Mach perceived that all questions concerning the basic model for the ultimate structure of matter were wide open. This was his reasoned opinion, in spite of the fact that he recognized the promise and plausibility of the atomic theory in explaining many facts. Unfortunately there were several different atomic conceptions to draw upon. Mach struggled with all of them. In fact, he invoked various atomic models to suit his intentions – just as everyone else did at the time.

Mach's philosophy of the role of the atomic theory was spelled out at some length in the general introductory remarks under the heading 'Methods of research'. Scientific activity, he indicated, might be envisioned as a three-stage operation. It begins with the discovery (by observation, experimentation, and induction) of the natural laws (*Naturgesetze*) which summarize the relationships that are seen to hold between natural phenomena. The discovery of many such laws (relationships) leads to the desirability of discerning fewer, and preferentially simpler, basic laws (*Grundgesetze*) by means of explanation (*Erklärung*), i.e., deduction. The basic laws are established by tying together natural laws and incorporating them into the existing structure of science. But there are fundamentally two kinds of explanation for Mach: "Either we detect that a given natural law is no more than the concurrence of several already known laws, or we search for intermediate links in going from one to the other."[31]

Mach's example of the former was the explanation of how central-force motion results from the combination of the laws of inertia and accelerated motion. An example of the latter was the explanation of how heat results from friction. "Experience teaches that a rubbed body becomes warm. If we now remark that friction generates oscillation in the molecules of the body, and that a certain kind of oscillation is heat, then we have inserted a middle term between the cause (friction) and the effect (heat), [namely] the oscillation of the molecules."[32]

Mach conceived hypothesis construction to be an indispensible concomitant of scientific activity. That is: "If one is not in a position to state the real cause of a phenomenon, one tries to invent such a one; one makes an hypothesis; one tests whether the phenomenon can be explained on the basis of certain presuppositions. Thereby one is generally guided closer to the goal." Furthermore, it is specified by Mach that where several hypotheses (with different presuppositions) are tested against the phenomena, that particular one will be accepted as the most persuasive and most rational which, in the least contrived way, explains the facts most completely and directly. But a single significant experiment which is not covered by the hypothesis can cause the theory to tumble: "Thus the corpuscular theory [of light] fails because it cannot explain interference." Usually, however, the decision between several theories is not so unambiguous and clear, and one tries to predict phenomena which will subject a given theory to an *experimentum crucis*. Mach's implication is clear: The atomic theory had not yet been subjected to such a test.[33]

The goal of physics, wrote Mach, is directed toward the search for the smallest number of highest principles (*oberste Principien*) which will explain the phenomena of nature. Before such highest principles are ascertained, scientists have no alternative but to accept the most convincing hypothesis available. Or, to quote Mach directly: "Now these highest principles have not yet been found [and] we must for the time being be content with the so-called atomic theory."[34]

Considerable emphasis was placed by Mach on the potency of simplicity inherent in atomic considerations applied to the study of equilibrium, motion, and the mechanical interactions resulting from the interplay of various forces. He recognized two kinds of atoms, *viz.*, atoms of matter (*Körperatome*) attracting each other as the inverse square of the distance of separation; and atoms of aether (*Aetheratome*) repelling each other with a force which falls off more rapidly than the square of the distance increases. The forces of attraction between atoms of matter and aether were taken to diminish more rapidly with distance than the forces of attraction between atoms of matter.[35]

The model of matter which Mach presented – 'an illustrative picture' – consisted of one or more spherical atoms of matter conglomerated by aether atoms. The latter were taken to be dense in the immediate neighborhood of the matter atoms, less dense away from the center, and thereafter rapidly reaching the normal density of aether atoms in free space (*im Weltraum*). To explain chemical differences Mach suggested that atoms of different kinds of matter be conceived as identical units whose individual properties depended upon their modes of grouping. The other most palpable alternative was to postulate inherent physical differences in the atoms of all the different chemical substances. Mach preferred the first of these alternatives.

Although Mach referred throughout the *Compendium* to matter-atoms and aether-atoms, he wanted to leave open the question of their fundamental common nature. Thus he wrote:

> Whether or not it is necessary to assume a special difference between aether and matter atoms can not be ascertained precisely at this time. It seems to me as if the assumption of one single kind of atoms, singly as in the aether atoms, and in larger groups as combined in matter atoms, just as well represents the phenomena. It would all depend only upon the form of the law of reciprocal action attributed to them. Nevertheless this view is not widespread and therefore we shall provisionally hold to the former view, which, in fact, since it is the more general one, at all times can be specialized into the second. The simplest hypotheses are always the most probable. On the other hand, a too severe specialization of the assumptions in natural science could cut off the way to one's goal so that the assumptions no longer can be modified during the process of investigation.[36]

On the above mentioned model of matter-aether atoms and aether-in-space atoms (possessing different modes of vibration and propagation depending on their arrangement and proportion) Mach attempted to supply plausible explanations for most of the well established physical properties of matter. Mach wrote:

Therefore, according to the atomic theory, all physical phenomena can be reduced to equilibrium and the motion of atoms. Physics would become applied mechanics, and therefore applied mathematics, if only we knew the law by which the atoms of matter and of aether act upon one another. Concerning that, however, we have at present only conjectures which pretty much reduce to what has been said above.[37]

In his *Compendium* Mach discussed in some detail the various physical phenomena which, he felt, could be analyzed more convincingly on the atomic than on the dynamic or continuous theory of matter. He wrote: "Although we do not know exactly the laws of acting forces between atoms, general conclusions can nevertheless be drawn from general presuppositions." Using the conventional model of molecular forces of attraction operating over "immeasurably small distances", at various discontinuous boundaries, Mach mathematically derived, as Laplace and others had done earlier, the conditions of equilibrium of fluid surfaces, capillary phenomena, and the shape of falling drops.[38] Compression, elasticity, thermal expansion, pressure-volume work, changes of state, melting, vaporization, the lattice forms and anisotropic characteristics of crystals were all 'explained' with the aid of the mechanical theory of heat and Mach's model of matter atoms immersed in aether atoms.

Especially noteworthy is Mach's discussion of the principle of the conservation of energy based on the mechanical theory of heat (which he later treated with such great derision). In evidence here are all the usual mechanistically-structured, atomistically-reasoned arguments for the mechanical equivalent of heat, the interconversions of kinetic energy (*lebendige Kraft*) and potential energy (*Spannkraft*), and the energy transformations characterized by chemical reactions and galvanic cells. All of physics thus becomes "nothing but applied mechanics" resting partly upon facts and partly upon hypotheses.[39] The properties of gases, too, were treated on the patterns of the kinetic-molecular theory of Krönig and Clausius.[40] The kinetic theory of gases was all but ignored by Mach in later years.

Chemical phenomena likewise were reduced to preferential mechanical configurations, motions, and interchanges resulting from the relative differences of attractive forces, or chemical affinities as chemists would have called them.[41] Chemical combinations accompanied by the release of heat were related to the attractive forces between components which give rise to particles with increased velocities.[42] The vigorous vibration of atom groups leading to chemical decomposition were accompanied by the release of the heat of vibration.[43] In view of the state of thermochemical investigations at this time, these views, we may add, are singularly unrefined mechanistic conceptions of the phenomena associated with heats of reaction.

Mach discoursed at some length on the kinetic-molecular theory of heat. Light he said was produced by heat, and heat by light; but Mach believed that "even if there is agreement that the movement of smallest particles is essential for heat phenomena, there still are the most diverse views concerning the nature of this movement". The phenomena of sound were equated with the movement of "systems composed of groups of aether atoms and matter atoms". The translational motion of atom groups accounted for the kinetic behavior of gases. The vibrations and axial rotations of the atoms were invoked to account for changes of state. Mach wrote:

It is most likely that all of these ways of movement occur simultaneously in most cases, each having some share in the heat phenomena. Indeed it would be difficult to comprehend how one of these movements could establish itself without, as it were, drawing the others along, generating them and changing over into them. ... Although it cannot yet be specified what heat is, one is nevertheless in a position to state whether a larger or smaller quantity of heat is present.[44]

The production of heat, said Mach, accompanies practically all physical phenomena, and is explained adequately by the kinetic or vibration theory of heat. For example, the phenomena of sound were attributed to the vibration of molecules or parcels of molecules involving however no internal spatial alterations of their constituent atoms. Now Mach took atoms to be hypothetical but he argued that there was nothing hypothetical about postulating smaller particles whose attractions follow the same law of gravitation which experience teaches for ordinary masses. He wrote:

Similarly the theory of sound contains no hypothesis, because the laws of elasticity, and many others upon which it is founded, are facts. By contrast, we base [our considerations] upon hypotheses when we reduce the phenomena of heat and light to the motions of atoms and molecules (atom groups), because it is merely an assumption, although a very probable one, that such atoms exist.[45]

A humorous note accompanied Mach's discussion of the atomic explanation for the transformation of the energy of sound into heat:

If Redtenbacher (in his dynamid system) reckons it unlikely that sound can be converted into heat, because it is impossible to heat a concert hall with a Bach fugue, then it is conceivable that one readily might be convinced to the contrary by listening to a composition of R. Wagner.[46]

Foremost among the phenomena of physics which were supported by the atomic hypothesis of matter were light and heat. The two were seen to be intimately related by Mach. Thus, he was thoroughly convinced that the mathematical analysis of the phenomenon of polarized light required a medium of discontinuous particles (aether atoms) in order to carry the transverse vibrations. He also concluded that the atomic hypothesis, more adequately than the dynamic theory, explained color refraction. In this case he reasoned that different colors and wavelengths could only be transmitted through one and the same medium with different velocities by assuming that the medium consists of discontinuous particles whose distances of separation are not vanishingly small compared to the wavelengths of the light. "The undulatory theory is a vital question (*Lebensfrage*) for physics, but the undulatory theory rests upon the atomic theory. Therefore atomistics is also a vital question for physics."[47]

Mach explained that light as a transverse vibration of aether atoms was propagated variously in matter and in free space – giving, for example, a manifold of possible modifications of reflection, refraction, etc. Light as motion was seen to produce the motion of heat in bodies, just as the motion of impact or friction produces heat. "The heat of bodies probably depends upon the standing vibrations of the aether between the atoms of matter; but perhaps also upon the vibration of the atoms of matter themselves."[48] One fundamental difference between light and sound was that light could be propagated as a vibration of aether atoms in matter-free space, whereas the propagation of sound was known to be impossible in the absence of matter atoms.[49]

Basic for Mach was the premise that light is a wave motion of the aether. He wrote:

If now the aether, in which a body (a system of atom groups) is immersed, is set into vibration, nothing is more natural than that the individual atoms which compose the body also are set into motion so that the illuminated body becomes warm. ... It might be said that heat is excited in a body in very much the same manner in which a stretched string is brought to vibrate simultaneously in resonance with a tone which is propagated through the air.[50]

Again, Mach attributed the cause of visible heat to the vibrating aether atoms which are associated with the system which constitutes a body immersed in aether (*ein in Aether eingetauchten Systems eines Körpers*).[51] The color, temperature, and frequency of light were all taken to be related to the mechanical vibrations of the aether atoms. But Mach was quite at a loss to explain why the periods of vibration should not be independent of the amplitudes of vibration as they are for a vibrating mechanical pendulum. "The forces here must operate according to other laws, if as it seems, the basic presuppositions actually are correct."[52]

As for the spectral investigations of Plücker, Kirchhoff and Bunsen, Mach conceived of the atoms "moving confusedly without taking any notice of each other". The most convincing relationship between vibrating atoms and their spectral lines, in accord with the most popular view of the day, was that of an acoustical analogy with vibrating strings giving 'one simple color' for each different spatial configuration of the atom in the atom grouping, but complicated by the possibility of vibration of the aether atoms in the immediate neighborhood.[53]

When he came to the problems of electricity and magnetism Mach discovered the atomistic conceptions to be somewhat remote and inapplicable. Still in an indirect way, he saw that it might be possible to reduce the phenomena to the motion of particles. Mach wrote:

> It is not yet agreed upon whether electricity and magnetism also are a certain kind of molecular motion; or perhaps whether, as was believed earlier, they are caused by the aggregation of certain fluids. In any case, electricity and magnetism are associated with molecular motions, because electricity can produce light and heat, i.e., vibrations. Vice versa, light, heat, and mechanical work can be transformed into electricity.[54]

Although Mach felt that he was successful, in part, in relating electricity to atomistic conceptions with the help of the energy conservation principle and the mechanical theory of heat, he otherwise perceived that "a deep darkness covers the nature of electricity".[55] One conceivable model was to suppose that the atoms were connected by elastic springs which become stretched on electrification and then are released in performing work. But this view presented many difficulties. For example: "It is not easy to explain on this [model] how two bodies which otherwise exert no noticeable action upon one another, can noticeably attract or repel one another as soon as they are electrified." Other models which had been proposed in the literature were also found to be unsatisfactory. The known thermoelectric effects afforded further complications which Mach could not readily bring into any atomistic scheme.[56]

An interesting incidental note tacked on to the end of the chapter on light reveals a line of thought which Mach explored at considerable length in later years. We quote:

> Perhaps it will be permitted to add a remark which strictly speaking does not belong here. If Fechner's view is correct, namely, that the psychic phenomena basically are only physical processes seen from another side, then we may hope to learn to know the most delicate processes in a special system of atoms (in our nerves) through the study of the sensations, and by this means we shall approach physics from another side. Temporarily, at least, I do not see any way to resist Fechner's assumption, least of all if one does not forget that it is a mere hypothesis.[57]

Mach's *Compendium* concluded with a note of caution:

> To the extent that it is possible by elementary means, we now have learned to recognize a series of physical phenomena and laws conceived from the point of view of mechanics and the atomic theory. The complete and rigorous reduction of physical laws to a few exact principles is nowadays not yet possible. The gaps are still too great. For heat and light phenomena we have at least found the general outline of a theory, and these outlines at least rest upon solid foundations. For electricity and magnetism we have not yet gained such insights. Instead of a true theory in these areas, there are many empirical rules which are necessary in order to find one's way among the greater number of facts. But precisely therein lies the enticement for further research; and we hope that here too we shall achieve the goal – to be sure, perhaps, only after a complete transformation of some of our basic physical conceptions. At least the peculiarities of electrical and magnetic behavior seem to point to the necessity of such a reform. ... The exact sciences attempt to reduce everything to the smallest possible number of laws and only thereafter search for new principles when the old [ones] no longer suffice in any way.[58]

We have seen that between the years of 1860 and 1863 Mach's language about and reference to atoms and molecules followed a highly mechanistic approach. He had not presented any significantly novel arguments in these works. His opinions, that is, did not deviate radically from those of most of his atomistically convinced contemporaries.[59] They all, more or less, treated the atom as a hypothetical entity. Mach's exploitation of the atomic theory for the description of natural phenomena was therefore comprehensive, systematic and impressive but fairly conventional. He had mastered the arguments of the atomic theory very well. That is to say, Mach's atomic deliberations, at minimum, demonstrate that although he later denounced the atomic theory vehemently and resolutely, he manifestly was not engaging in such polemics because of blindness or ignorance of the strength and weakness of these conceptions.

Having said this much, it is necessary to add that Mach's own experimental investigations gave him no convincing support for the atomic theory. He was obviously looking for such support. It is true that his studies were scarcely extensive enough to merit strong changes of opinion. I suggest that part of the reason for Mach's eventual rejection of atomism is related to the fact that he was engaged in and interested in problems for which the atomistic-mechanistic conceptions of physics were of little value, or else, in which the mathematics was inherently complex enough to cover up any underlying physical clues about the nature of matter.

Mach applied himself to the problem of the shape and cohesion of fluids, capillarity, and their molecular forces – all of which are now reducible to electrodynamic considerations without reference to Newtonian attractive forces between particles. Similarly, Mach's problem of the mechanism of transmission of mechanical impulse in systems of contiguous particles lends itself nicely to wave equations designed for continuous media. Mach was interested in the Doppler effect, pulse studies, acoustics, resonance, and the transmission of sound in the ear. These can be subjected to Fourier analysis without any concern about the nature of matter. Mach's study of discontinuous line spectra in relation to chemical differences provided the most promising hope of seeing chemistry transformed into applied mechanics. But the historical developments which led from spectra to information concerning the internally structured atom turned out to be a mare's nest of confusion. How could a simple atom of incandescent iron possibly show so many spectral lines – even if its modes of vibration were analogous to the tones and harmonics of a bar or sphere in a state of violent vibration?

Mach simply was not engaged in the kind of research that might have added support to the atomic theory. If he had been working, as some chemists were, on the problem of fixing chemical composition or on a method for establishing unambiguous atomic weights and formulas, then Mach might not have broken faith with the atomic theory. Or, if he had been engaged, as Clausius, Stefan and Puschl were, in studies on the kinetic theory of gases, or the conduction of heat in gases, or the specific heats of gases and solids, Mach very likely would have been more tolerant about atomistic conceptions. In fact, he was slowly shifting over in his scientific interests from physics to the physiology and psychology of sensation, *viz.*, the new discipline of psychophysics.

How this new shift of orientation in Mach's scientific interests influenced his conceptions about mechanism and atomism began to make itself known in some lectures on psychophysics which were delivered at the University of Vienna on Saturdays and Sundays during the summer of 1863.[60]

The dominant theme in these lectures was to demonstrate that the expression 'exact sciences' did not necessarily characterize any particular species of science but rather the stage of development of any science. To give a specific example: "Psychology, the science of the phenomena of mental life (*des Seelenlebens*), and psychophysics, the science of the relation of psychological and physiological phenomena, are capable of completely becoming exact sciences, provided they merely continue further along their trodden path."[61]

What did Mach mean by applying 'exact laws' to the study of mental life? He was searching for an answer to this question, not merely by trying to reduce psychophysics to applied mechanics or atomistic considerations or the law of conservation of energy – much as he felt that the "belief in the correctness of these basic hypotheses was increasing day by day". Rather, he wrote:

> We must and can go another step further. Man too is controlled by exact laws, which of course are not revealed easily in individuals; but therefore all the more distinctly in the larger whole. To demonstrate these laws we do not wish to proceed deductively from the much lauded materialistic proposition that all phenomena, even the intellectual [ones], rest upon material processes, and that they therefore necessarily are determined mathematically. Rather we want to avoid all hypotheses [and] strike out on a rigorous path of experience, such as was trodden by the statisticians, especially Quételet.[62]

Of great importance for the study of psychophysical phenomena in man, Mach indicated, were the developments in probability theory, and the analysis of observational errors, as applied to moral and intellectual questions. The new researches of Wundt, Fechner, and Helmholtz and the psychology of Herbart, Mach indicated, were supplying the close connection between psychophysics, pathology, psychology, physiology, and the statistics and mechanics of the mind (*Geist*). "I avoid", said Mach, "all metaphysical questions."[63]

"What the soul is", wrote Mach, "that is as much a matter of indifference for mathematical psychology as the essence of matter [*Wesen der Materie*] is for physics. Ideas [*Vorstellungen*] can be thought of as a state [*Zustand*] of the soul." To say with Herbart that ideas persist after the stimulus of the perception has ceased is analogous to the law of the inertia of matter. "Indeed, there might even be a deeper connection between these two propositions." In Herbart's study of ideas we encounter meaningful questions pertaining to: their mass and equilibrium, points of minimum arrest, and the square of the intensity of what survives in consciousness. Said Mach: "These cases, which Herbart treats, are simple ones which never occur in reality; but this is necessary for the beginning of an investigation." Mechanics proceeds on similar elementary assumptions. "For the good self-observer it is more than a mere picture that the ideas reciprocally crowd one another. We, so to speak, really feel this inside ourselves."[64]

Mach moreover considered some questions related to stimulus and sensation in connection with the Fechner-Weber law. Far from expecting to reduce 'stimulus' and 'sensation' to some kind of unitary atomic entity, Mach suggested that they be looked upon as heterogeneous magnitudes as different as space and time in physics. Beyond that, Mach believed that the sensations of light, sound, and gravity should be measured as fundamentally different quantities. He said that we simply cannot throw everything, as physics does, into "the three pails which are labelled mass, time, and space".[65]

Mach's lectures in psychophysics included discussions on the sensations of feeling, color and tone, as well as inquiries in music theory in relation to the ear and its pathology. In the discussion of the Young-Fechner-Helmholtz three color theory of sensation Mach remarked that the atomic interpretation – three kinds of atoms constituting the nerve fibers in the retinal receptors – might just as well be reduced to anatomical or chemical differences in the retina without the hypothesis of atoms.[66]

In discussing the well-known phenomena of tone and color sensation – according to which sensation is related to the stimulus of specific nerves and not the nature of the stimulus – Mach questioned the meaning of the theory (held by some scientists) that all such phenomena can be reduced to the flow of electric currents. Such a theory, said Mach, says nothing about the processes in question. It corresponds with a pertinent remark once made by Ludwig:

If we know the chemical elements of the nerves, we know no more about their nature than we know about a telescope, if we are told that it consists of glass, metal and lampblack. The electrical current only shows us the outside of nerve processes.[67]

Not at all hopeful of being able to reduce psychophysical phenomena to atomic considerations, Mach nevertheless in his last lecture on psychophysics, concluded that there is really nothing to take the place of the atomic theory as a temporary hypothesis when it comes to the treatment of man's thought (*Gedanken*) process. Repeating a famous materialistic proposition, Mach wrote: "Thought relates to the brain, as gall to the liver, and urine to the kidneys" – a very 'uropoetic' way of accounting for the origin of thought. This expresses what Mach believed to represent the very close and real connection between psychic and physiological processes. Nevertheless:

It takes no deep thinking to discover the lame aspect of the simile. Thought is no secretion. No one has yet discovered a thought-fistula as one finds urinary and gall fistula. One cannot say that thoughts are physical processes in the brain. No matter how carefully we investigate physical processes, we discover molecular motions, it is true, but no thoughts. The electrical currents in the brain are, nevertheless, essentially something else than the conception of green color which the investigated subject perhaps even possesses. It is more likely our conceptions that we discover while investigating the brain and those of the investigated brain.[68]

For Mach it was simply unnecessary to adopt an hypothesis concerning the nature of the 'psychic' while investigating the facts of experience. The progress of science, Mach thought, did not demand any over-hasty moves at this point. The connection between the psychic and the physical was real although the boundary conditions between them were unperceived. But how was their connection to be ascertained? Perhaps, remarked Mach, they are really the same processes perceived differently, being analogous to Fechner's convex exterior and concave interior of the same natural phenomena. Most persons materialize the problem, wrote Mach, but that view has inherent difficulties. It is only certain processes such as the transverse vibration of light and the spectra of incandescent gases which are best accounted for on the atomic theory.

Spectral analysis, accordingly, to use a familiar medical model, suitably could be called the percussion and auscultation of the atoms. If we want to intermix as few hypotheses as possible, then we must say that physics always and everywhere leads to the assumption that there are a finite number of essences and forces active in phenomena.

Mach inquired: "How are we supposed to think about atoms? Colored, glowing, sounding, hard?" These are not the properties of the atoms but the properties of matter consisting of atoms. Might the atoms then be conceived merely as centers of force. The answer is "no"! For what would be the meaning of saying that one center of force acts upon another? Mach wrote:

In short let us make a confession. We can in no way, rationally speaking, claim an exterior [*Aussenseite*] for the atom. If we must conceive of anything we must attribute an interior [*Innenseite*] to it; an inner nature [*Innerlichkeit*] in some measure analogous to our soul. In fact, how should the soul suddenly come to be a combination of atoms in the organism, if the seed did not already reside in the individual atom?

The problem is to find in the organism those physical processes whose reverse sides are the psychic processes; i.e., "Where is the seat of the soul to be found?"[69]

Mach suggested, finally, that it is a worthy goal for science to strive for the reduction of many riddles to one riddle. The riddle of the atomic hypothesis might thus conceivably qualify as that one riddle. The test of a theory, Mach felt, was not so much in joining together all the facts as in discovering new facts explained exclusively by that theory. "For the value of an hypothesis consists mainly therein, that by a kind of *regula falsi* it always leads closer and closer to the truth."[70] For Mach the atomic theory was then after all at least a kind of *regula falsi.*

In 1864 Mach moved to Graz to accept a full professorship in mathematics at the university there. For the first time in his life the salary was adequate. Although he was permitted to lecture on physics, he was given neither an institute nor money for scientific apparatus. He therefore drew on his own private funds to secure the equipment which he needed for his own research. By 1866 he was given the title of Professor of Physics at Graz, but he continued to pursue the physiological and psychophysical problems which already had captured his imagination before he left Vienna.

By 1864 Mach's views on the atomic theory were under serious revision. In that year he published a second short note on the spectra of incandescent gases, again in Schlömilch's journal.[71] He knew that several physicists had experienced displeasure on reading his note on chemical spectra.[72] They had said his views were "*zu naturphilosophisch*". He replied: "In response I must remind [you] that I by no means consider the atomic theory as something established in itself, but rather as a useful, temporary empirical formula; as a kind of *regula falsi*, to be used in drawing closer to the truth."

Mach's main argument in this note reminds us of a statement in the preface to his *Compendium der Physik*, *viz.*, that the results acquired with the help of the atomic theory can always be "translated into the concepts of another metaphysical view of matter which, henceforth, will prove to be more cogent". And to this he added: "I personally am now already very much inclined toward such a translation. And I shortly hope to show that my remarks by no means hang together so necessarily with theories, but can be understood very readily as an expression of the facts."[73]

The information from gaseous bright line spectra, Mach reasoned, provided a temporary empirical formula (a kind of contrived or false rule), which made use of the number and relative spatial configurations but not the 'nature' of the atoms. Mach knew that his evidence for the scheme was rather meager and so he offered an apologetic and explanatory rationale for his theory:

> Naturally I have only drawn my remarks from the calculation of a single example, and since then decided upon a general calculus which partly gave other results, [and] which I hasten to add here as a correction and supplement.
>
> 1. A gas molecule consisting of *n*-atoms, put into concussion [*Erschütterung*], gives 3*n*-6 different modes of vibration [and] a spectrum of 3*n*-6 bright lines, which are situated just like the atoms. For molecules of less than 3 atoms the formula no longer holds. A molecule consisting of 2 atoms gives [only] one mode of vibration. The color but not the number of the [spectral] lines depends upon the [relative] location of the atoms.
>
> 2. If heat and light depend on the vibrations of atoms then Kirchhoff's law – concerning the equality of absorptive and emission power for all kinds of rays at the same temperature – obviously follows from the laws of resonance.
>
> 3. Where the resistances to which the vibrations of the atoms are exposed are small, the molecule will also most readily receive those vibrations which it is able in a self-luminous way to emit.
>
> Moreover, the assumption of a finite number of atoms composing a gas molecule agrees well with many other phenomena to which I will return later.[74]

To cite an example of Mach's progressive disappointment with atomistic and mechanistic explanations, there is a paper, which he delivered in 1864[75], which takes up the discussion of problems in physiological acoustics where he had left them in two previous papers of 1863.[76] The work includes numerous comments about his earlier experiments on the auditory organ, the pulse, the Doppler effect, and the spectral study of gases. Still, there is no mention of the kind of atomic considerations with which he embellished all of these matters in his papers a few years earlier. Indeed the entire paper reads like a roster of negative *physical* results and *physical* explanations. To wit – the following typical comments are scattered throughout the paper: Concerning the change of dynamics of aural tones in relation to the rhythm of the pulse – "I have not succeeded yet with an explanation."[77] In relation to the elevation of pitch as a function of distance, or 'subjective diminution' Mach said: "First of all we are certain that the pitch really cannot be altered, since we understand pitch as frequency of vibration. The elevation therefore is only subjective, apparent, physiological ... . The question now is another one. From whence comes the disappointment? Or, to speak more accurately, from where comes the physiological phenomenon, since we no longer are dealing with a physical one."[78]

It is not our intention here to analyze the criticisms of the atomic theory which appeared in Mach's papers and monographs between 1864 and the year of his death in 1916. Nevertheless, it is appropriate for us to consider what Mach had to say in retrospect concerning the period of his life which we have treated above, *viz.* before he categorically denounced the atomic theory.

In his first critical and historical monograph (*History and Root of the Principle of Conservation of Energy* – 1872)[79] Mach tells us how he "was led to the view that we need not necessarily represent to ourselves molecular-processes [*Molecularvorgänge*] spatially, at least not in the space of three dimensions".

In the year 1862, I drew up a compendium of physics for medical men, in which, because I strove after a certain philosophical satisfaction, I carried out rigorously the mechanical atomic theory. This work first made me conscious of the insufficiency of this theory, and this was clearly expressed in the preface and at the end of the book, where I spoke of a total reformation of our views on the foundations of physics.

I was busied, at the same time, with psychophysics and with Herbart's works, and so I became convinced that the intuition of space is bound up with the organization of the senses, and, consequently, that we are not justified in ascribing spatial properties to things which are not perceived by the senses. In my lectures on psychophysics, I already stated clearly that we are not justified in thinking of atoms spatially. Also, in my theory of the organ of hearing, I brought before my readers the series of tones as an analogue of space of one dimension. At the same time the quite arbitrary and, on this account, faulty limitation of the number of dimensions in Herbart's derivation of "intelligible" space struck me. By that, now, it became clear to me that, for the understanding, relations like those of space, and of any number of dimensions, are thinkable.

My attempts to explain mechanically the spectra of the chemical elements and the divergence of the theory with experience strengthened my view that we must not represent to ourselves the chemical elements in a space of three dimensions. I did not venture, however, to speak of this candidly before orthodox physicists. My notices in Schlömilch's *Zeitschrift* of 1863 and 1864 contained only an indication of it.[80]

The other place in which Mach reminisced about his early atomistic years was in the third of three papers (1903) originally published in *The Monist*, which was edited by Paul Carus of the Open Court publishing company in Chicago.[81]

In this paper Mach examined some concepts in physics in relation to space and the geometry of the objects of sensuous experience. For "mere things of thought like atoms and molecules, which from their nature *can never be made the objects of sensuous contemplation*" Mach felt no obligation to think in terms of three-dimensional Euclidean relationships. He wrote:

While still an upholder of the atomic theory, I sought to explain the line-spectra of gases by the vibrations of the atomic constituents of a gas-molecule with respect to another. The difficulties which I here encountered suggested to me (1863) the idea that non-sensuous things did not necessarily have to be pictured in our sensuous space of three dimensions. In this way I also lighted upon analogues of spaces of different numbers of dimensions.[82]

If we were to examine Mach's scientific papers in the decade after 1863/4 we would note, in particular, that there is a marked increase in experiments dealing with the psychology and physiology of sensation – and this without any noteworthy concern with atomic conceptions. For approximately ten years atoms are virtually dismissed in Mach's deliberations, at least in print. The new interest in psychophysics included: study of the physiology of musical acoustics; the sensation effects of various experimental tones on the ear; the accommodation, and the spatial and time sensation of the eye and ear; visual sensation and excitation of the retina, etc. In all of this work it is obvious that Mach was not at all restricting himself to physics, nor, specifically to mechanistic or physical interpretations of physiological and psychological phenomena.

Mach, however, most certainly had not abandoned physics. Concurrent with his new psychophysical interests he had been experimenting with and writing about such matters as suspended particles in fluids, wave and vibration phenomena, photography, stereoscopy, stroboscopic effects, and music theory. More important still, Mach had been deeply engaged in thinking about the foundations of space, time, mass, mechanics and thermodynamics.

Mach's new views on space and time were first communicated to the public in his lectures on the principles of mechanics at the University of Vienna in the summer semester of 1862, as well as in the second round of his lectures on psychophysics, namely, at Graz in the winter semester of 1864/5. Brief reference was made to some of his views on space in 1865 and 1866[83], when he was exploring concepts of space and geometry other than the 3-dimensional Euclidean space which describes physical objects in physical space. These discussions dealt with the space of sight, touch and muscles: Sehraum, Hautraum and Muskelraum. There was not a word about atoms.

The criticism of atomism proper, did not come prominently to the fore until 1872 in Mach's *Geschichte und Wurzel.* This fiery little monograph can be looked upon as the incipit for the second distinct phase of Mach's views regarding the ultimate structure of matter. In that work Mach's views on the atom more explicitly were tied in with his larger conception of the role of hypothesis and fact in theory construction.

T. H. Huxley is supposed to have said that a great tragedy in science is the slaying of a beautiful hypothesis by an ugly fact. Mach probably would have said: The great blessing of science is the slaying of an ugly hypothesis by a beautiful fact.

Mach eventually conceived the existence of the atom to be impossible. Whenever in later life, in Vienna, anyone spoke about atoms in his presence he would interrupt with the question: Have you seen one? (*Habn S' eins gsehn*?) Conan Doyle recognized and used a good principle: whenever you have eliminated the impossible, whatever remains, however improbable, must be true. When Mach slew the atom there still remained the continuum. Did he ever seriously reckon with it? Or was he justified, in his day, in pushing aside such metaphysical issues in the belief that it suffices to consider the relationship of phenomena to one another? He had his own metaphysics to support him in that view.

University of Wisconsin

REFERENCES

* It is a pleasure to acknowledge the support of the National Science Foundation for research on the life and works of Ernst Mach.

[1] In another paper the author will examine Mach's criticism of the atomic theory as contained in his major treatises, papers, and correspondence with scientists and philosophers.

[2] Ernst Mach, 'Einleitung', in *Beiträge zur Doppler'schen Theorie der Ton und Farbenänderung durch Bewegung*, Prag 1873, p. 3.

[3] Christian Doppler, 'Ueber das farbige Licht der Doppelsterne und einiger anderer Gestirne des Himmels', *Abhandl. Böhmischen Ges. Wiss.* [5] **12** (1843), 467–482.

[4] Buijs Ballot, 'Akustische Versuche auf der Niederländischen Eisenbahn, nebst gelegentlichen Bemerkungen zur Theorie des Hrn. Prof. Doppler', *Ann. Physik* **66** (1845), 321–351.

[5] J. Petzval, 'Ueber ein allgemeines Princip der Undulationslehre: Gesetz der Erhaltung der Schwingungsdauer', *Wien, Ak. Sber.* **8** (1852), 134–156; 'Ueber die Unzukömmlichkeiten gewisser populärer Anschauungsweisen in der Undulationstheorie und ihre Unfähigkeit das Princip der Erhaltung der Schwingungsdauer zu ersetzen', *ibid.*, 567–586, and *ibid*, **9** (1852) 699–737; Doppler, 'Bemerkungen zu dem Aufsatze: Über ein allgemeines Princip der Undulationslehre, usw.', *ibid.*, 587–593; v. Ettingshausen, 'Bemerkung, derselben Gegenstand betreffend', *ibid.*, 593–594.

[6] Ångström, 'Optische Untersuchungen', *Ann. Physik* **170** (1855), 141–165.

[7] Ernst Mach, 'Ueber die Änderung des Tones und der Farbe durch Bewegung', *Wien, Ak. Sber.* **41** (1860), 543–560; continued in *Ann. Physik* **116** (1862), 333–338.

[8] This was first undertaken by A. Belopolsky with an arrangement of two counter-rotating discs with mirrors moving at the rate of 80 meters per second. He measured the shift in the solar Fraunhofer lines. See *Astrophys. J.* **13** (1901), 15–24. Several years later B. Galitzin and J. Wilip measured the Doppler shift spectroscopically using a mercury lamp source and a Michelson echelon. See *St. Pét. Ac. Sc., Bull.* [VI] **1** (1907), 213–223.

[9] Ernst Mach, *op. cit.* (1860), p. 548.

[10] Ernst Mach, 'Ueber die Controverse zwischen Doppler und Petzval, bezüglich der Änderung des Tones und der Farbe durch Bewegung', *Z. Math. Physik* **60** (1861), 120–126.

[11] *Ibid.*, p. 124.

[12] Ernst Mach, 'Ueber die Molekularwirkung der Flüssigkeiten', *Sber. Akad. Wiss. Wien* **46** (1863), 125–134.

[13] *Ibid.*, p. 127.

[14] *Ibid.*, pp. 132–133.

[15] Ernst Mach, *Popular Scientific Lectures*, La Salle, Illinois (1943), 1–16.

[16] *Ibid.*, p. 10.

[17] *Ibid.*, pp. 15–16.

[18] Ernst Mach, 'Ueber die Spectra chemisch verschiedener Körper', *Z. Math. Physik* **17** (1862), 214–216.

[19] *Ibid.*, p. 216.

[20] Ernst Mach, 'Ueber die Gesetze des Mitschwingens', *Sber. Akad. Wiss. Wien* **47** (1863), 33–48.

[21] *Ibid.*, p. 35.

[22] Ernst Mach, 'Zur Theorie des Gehörorgans', *Sber. Akad. Wiss. Wien* **48** (1863), 283–300.
[23] *Ibid.*, p. 284.
[24] *Ibid.*, p. 290.
[25] *Ibid.*, p. 292.
[26] *Ibid.*, p. 300.
[27] Ernst Mach, 'Die Leitgedanken meiner naturwissenschaftlichen Erkenntnislehre und ihre Aufnahme durch die Zeitgenossen', *Scientia* **7** (1910), 234.
[28] Ernst Mach, *Compendium der Physik für Mediziner*, Vienna 1863.
[29] *Ibid.*, p. vi.
[30] *Ibid.*, pp. 3–4.
[31] *Ibid.*, p. 10.
[32] *Ibid.*, p. 10.
[33] *Ibid.*, pp. 10–11.
[34] *Ibid.*, p. 11.
[35] *Ibid.*, p. 15.
[36] *Ibid.*, p. 188.
[37] *Ibid.*, p. 18.
[38] *Ibid.*, p. 18. Also pp. 76–85 and 89–91.
[39] *Ibid.*, pp. 49–55.
[40] *Ibid.*, pp. 91–96.
[41] *Ibid.*, pp. 16–18.
[42] *Ibid.*, p. 171.
[43] *Ibid.*, p. 172.
[44] *Ibid.*, pp. 161–163.
[45] *Ibid.*, pp. 55–56.
[46] *Ibid.*, p. 170.
[47] *Ibid.*, p. 14.
[48] *Ibid.*, p. 16. See also pp. 178–179 and 189–193.
[49] *Ibid.*, p. 187–188.
[50] *Ibid.*, p. 170.
[51] *Ibid.*, p. 178.
[52] *Ibid.*, p. 178.
[53] *Ibid.*, pp. 178–181.
[54] *Ibid.*, p. 16.
[55] *Ibid.*, p. 234.
[56] *Ibid.*, pp. 235–239.
[57] *Ibid.*, p. 234.
[58] *Ibid.*, pp. 271–272.
[59] The subject of Mach's use of atomistic models in relation to views on the ultimate structure of matter, as seen among his immediate predecessors and contemporaries, forms the basis of another paper.
[60] 'Aus Dr. Mach's Vorträgen über Psychophysik', *Oesterr. Zeitschrift für praktische Heilkunde* **9** (1863), 146–148, 167–170, 202–204, 225–228, 242–245, 260–261, 277–279, 294–298, 316–318, 335–338, 352–354, 362–366.
[61] *Ibid.*, p. 147.
[62] *Ibid.*, pp. 147–148.
[63] *Ibid.*, pp. 167–170.
[64] *Ibid.*, pp. 202–204.

[65] *Ibid.*, pp. 225–227, 242–245.
[66] *Ibid.*, p. 317.
[67] *Ibid.*, p. 335–336.
[68] *Ibid.*, pp. 362–363.
[69] *Ibid.*, p. 364.
[70] *Ibid.*, p. 366.
[71] Ernst Mach, 'Vorläufige Bemerkungen über das Licht glühender Gase', *Z. Math. Physik* **9** (1864), 69–70.
[72] *Supra*, pp. 82–83.
[73] *Ibid.*, p. 69, 1864.
[74] *Ibid.*, p. 69–70.
[75] Ernst Mach, 'Ueber einige der physiologischen Akustik angehörige Erscheinungen', *Wien, Ak. Sber.* **50** (1865), 342–362.
[76] *Supra*, pp. 83–85.
[77] *Ibid.*, p. 348.
[78] *Ibid.*, pp. 351–352.
[79] Ernst Mach, *Die Geschichte und Wurzel des Satzes von der Erhaltung der Energie*, 2nd unaltered reprint, Leipzig, 1909. The cited quotations and references are from the Jourdain translation, Open Court, Chicago, 1911.
[80] *Ibid.*, pp. 86–88.
[81] The three essays were published together in 1906 by Open Court as *Space and Geometry in the Light of Physiological, Psychological and Physical Inquiry*. The citation is from this edition.
[82] *Ibid.*, pp. 138–139.
[83] Ernst Mach, 'Bemerkungen zur Lehre vom räumlichen Sehen', *Z. für Phil. und philos. Kritik* **46** (1865), 1–5; 'Bemerkungen über die Entwicklung der Raumvorstellungen', *ibid.* **48** (1866), 227–232.

KARL MENGER

MATHEMATICAL IMPLICATIONS OF MACH'S IDEAS: POSITIVISTIC GEOMETRY, THE CLARIFICATION OF FUNCTIONAL CONNECTIONS

The highest tribute to a scientist on an anniversary of his death is the demonstration that his ideas are still alive. Fifty years after Mach, some thoughts of this great Austrian philosopher-scientist have an unspent vitality. Two ideas of his on which I wish to concentrate lead to implications for mathematics that have not yet been fully explored: Mach's general philosophical views suggest the development of a positivistic geometry, which has hardly begun; and his maxim that causal explanations should be replaced by functional connections calls for a clarification of the method of applying the function concept and other ideas of mathematical analysis to science – a clarification which is still in its initial stage.

I. POSITIVISTIC GEOMETRY

Euclid's *Elements* are an idealized description of the space in which we live and experiment. But Euclid's assumptions transcend the realm of the observable in two directions: in the large and in the small.

With regard to our space *in the large*, Euclid assumed the possibility of extending straight lines without ever coming to an end and without ever returning to the starting point. The former assumption is made in the *Elements* explicitly; the latter, tacitly. A further hypothesis about the space in the large, somewhat more complicated and less plausible, is the famous Fifth Postulate, about parallel lines, which is the foundation of the theory of similar figures and which accounts for the theorem about the sum of the angles in a triangle. While that axiom can be neither verified not falsified directly (i.e., by rods or even light rays) some of its consequences do lend themselves to tests by observation. One can, for instance, measure the angles in a triangle. So, when Bolyai and Lobachevsky challenged Euclid's postulate by using a different assumption as the basis of a non-euclidean geometry, Gauss actually did measure those angles, undisturbed by Kant's doctrine of the *a priori* necessity of Euclid's system. Obviously, Gauss considered geometry as the science of our physical

Boston Studies in the Philosophy of Science, VI.

space; and, in the spirit of positivism, all he claimed as the result of his measurements was a tentative confirmation of Euclid's theory: a confirmation within the limits of errors of observation and confined to terrestrial triangles, which are macroscopic but not megascopic figures.

Riemann developed Bolyai's, Lobachevsky's, and his own non-euclidean geometries into a general theory of space, which is differential-geometric in character. But the word *differential* should not mislead us into the belief that he modified Euclid's geometry in the small. Sufficiently small triangles in Riemann's geometry differ from Euclidean triangles as little as we please. Significant deviations from Euclidean geometry occur only in the large. This is why it took more than a century until these modifications of Euclid's geometry were applied to nature, namely, in connection with the general theory of relativity, and why, somewhat paradoxically, the applications of this differential geometry concern astronomical distances and cosmology.

Euclid's assumptions about space *in the small* may be said to be conspicuous by their absence. Because of the laws about similar figures, his theory simply amounts to an interpolation of the macrocosmic facts into microcosmic domains; and, for all practical purposes, the same can be said about the non-euclidean and Riemannian geometries. Perhaps because of their implicitness, these assumptions of Euclid have never been challenged as seriously as were his explicit postulates about space in the large.

It is in this respect that a positivistic approach calls for revisions, in fact, for changes that are much more profound than those that have led to Riemannian geometries. Looking at space in the small from a positivistic point of view we must question the very framework of Euclid's system and his most basic concepts.

The chapter on the continuum in Mach's *Prinzipien der Wärmelehre* contains the following passage: "All that *appears* to be a continuum might very well consist of *discrete* elements, provided they were sufficiently small compared with our smallest practically applicable units or sufficiently numerous." (Parenthetically, it may be pointed out that the quoted passage favorably compares with less fortunate assertions of Mach's in his polemic against atomism. Some of those assertions bordered on a dogmatism conflicting with Mach's own general philosophical ideas and subsequently were disproved by experiments. But it should also be emphasized that Heisenberg's scheme of confining atomic theory to the connection of observable quantities restored Mach's program; and that today's view of atomic physics as a method which, from present observations, derives probabilistic predictions about future observations is the very epitome of Mach's ideology. Yet, perhaps because of Mach's dogmatic fight against atomism, his programmatic views are rarely quoted in quantum mechanics – certainly not as frequently as in relativity theory.)

As far as geometry is concerned, the passage quoted from the *Wärmelehre* actually says more than is justified in a strictly positivistic geometry. With regard to the ultimate structure of space (whatever that may mean), a positivistic geometer should out-Mach Mach and refrain from making any statements whatsoever. The concepts of classical geometry that he must challenge include the notions of point, line, and congruency as well as the post-euclidean concept of numerical distances.

Since points and lines do not exist in the macrocosm one might at first think that these very concepts represented the idealizations of macrocosmic objects such as dots or cross hairs and rods or light rays that microgeometry requires. Actually, however, points and lines cannot be ascertained in the microcosm any more than in the macrocosm. How can one locate, identify, and recognize a point and determine straight segments (if in the small there is anything deserving this name!) or ascertain that a point and a line are incident and that three points are collinear?

It has often been said that, from an empirical point of view, the application of irrational numbers to space is illusory since no observation can distinguish irrational from rational coordinates. But neither is a rational number or even an integer ascertainable as the abscissa of a point on a line. Whereas in counting discrete elements, say apples in a basket, one may arrive at the number 12 and then is sure that the contents are not 11.97 or 12.02 apples, the statement that the abscissa of a point on a line is 12 cannot be more strongly supported than the claim that the abscissa be 12 plus some sufficiently small but otherwise arbitrarily chosen number, which may be rational or *irrational*. From the positivistic point of view, the difficulty in describing continua does not lie in the use of special kinds of numbers. It lies in the identification of the individual elements, in particular, in their description by numbers of any kind. This difficulty precludes of course a positivistic application of the limit concept. What we actually observe or immediately derive from observations are difference quotients and product sums, which can be treated by numerical methods. Whereas, in mathematical analysis, difference quotients and product sums are considered as approximations to derivatives and integrals, from a positivistic point of view, most statements involving derivatives and integrals are unverifiable approximations to assertions about difference quotients and product sums – but very convenient approximations, because calculus has a simple algebraic structure and is well developed.

No one was more keenly aware of the problems connected with the notion of the physical continuum than Poincaré, who emphasized the difficulties on many occasions, e.g., in *La science et l'hypothèse*, *La valeur de la science*, and *Des fondements de la géométrie*. In all these books he characterized the physical continuum by the formulae

$$(1) \qquad A=B, \quad B=C, \quad A \neq C,$$

symbolizing the fact that an element of a physical continuum may be indistinguishable from two others that can be distinguished from one another. The *mathematical* continuum, according to Poincaré, is a construct whose main purpose it is to overcome the formulae (1), which he called contradictory or at least repugnant. He finally returned to the problem in *Dernières Pensées*, where he proposed the following definition of a physical continuum. It is a system S of elements, any two of which can be joined by a chain of elements of S such that each element of the chain is indistinguishable from the next element.

While, after Poincaré, these ideas of his have not in my opinion been sufficiently developed, two steps have been taken in the direction of a positivistic geometry.

The first step consisted in inverting the traditional procedure of taking *point* as the primitive concept and of subsequently defining spatial objects as sets or unions of points. Whitehead and Nicod started with *lumps* or *pieces* and defined points as certain sets or intersections of lumps. Pursuing a course suggested by dimension theory, in the early 1930's Wald defined points of topological spaces as certain denumerable sequences of lumps. R. L. Moore applied similar ideas to the points in the plane. Somewhat related theories, which, however, were essentially oriented toward algebra rather than geometry, were developed by Stone, Wallman, and Milgram, who introduced points as (indenumerable) sets of elements of a Boolean ring, a distributive lattice, and a general partially ordered set, respectively. A brief summary of all these theories from a geometric point of view and a bibliography are contained in my lecture 'Topology without Points' in the *Rice Institute Pamphlets.*[1] But as that title indicates, these theories were confined to the topological aspects of geometry and did not include the ideas of congruency, distance and the like.

The second step was the development of a statistical or probabilistic metric geometry, which I outlined in the early 1940's.[2] Whereas traditionally one associates, with any two points p and q of a space, a number $d(p, q)$, called the distance between p and q, the probabilistic theory is based on the assumption that with any two elements a distribution function F_{pq} be associated. Its value for x, that is, the number $F_{pq}(x)$, may be interpreted as the probability that the distance between p and q be $\leqslant x$. From the fact that distances are certainly non-negative it follows that, if p and q are any two elements, then $F_{pq}(x)=0$ for all $x<0$. With increasing x, the values $F_{pq}(x)$ monotonically increase from 0 to 1. If $F_{pq}(x')=0$ for some $x'>0$, then the distance between p and q certainly exceeds x'. If $F_{pq}(x'')=1$ for some x'', then the distance between p and q certainly does not exceed x''. If $F_{pq}(x)<1$ for all x, then, with increasing x, the numbers $F_{pq}(x)$ asymptotically approach 1. A numerical distance $d(p, q)$ between p and q, as assumed in the classical theory, may be said to exist in the special case where $F_{pq}(x)=0$ if $x<d(p, q)$, and $F_{pq}(x)=1$ if $x>d(p, q)$. Normal distributions, on the other hand, are ruled out by the fact that $F_{pq}(x)=0$ for at least all negative x.

The symmetry of the distance is expressed by the assumption that, for each pair of elements, $F_{pq}=F_{qp}$; that is to say, $F_{pq}(x)=F_{qp}(x)$ for all x. Further assumptions are needed in order to reflect the basic geometric fact that for any three elements none of the three distances exceeds the sum of the other two. They can be expressed in inequalities connecting the distributions F_{pr} with the distributions F_{pq} and F_{qr}. Extensive studies of such triangle inequalities (one of which was proposed by Wald[3]) and of related subjects have been carried out during the past few years by Schweizer and Sklar[4], by Šerstnev[5] in Russia, and by others. A report by Schweizer on this new branch of geometry will appear in one of the volumes of *Probabilistic Methods in Applied Mathematics*, edited by A. T. Bharucha-Reid.[5a]

In contrast to many postulates in classical geometry, some assumptions about the elements of a probabilistic space are such as to be valid if $p, q, r, \ldots$ are thought of as points or as lumps or as certain other entities.

If p and q are lumps, then one may interpret $F_{pq}(x)$ as the probability that the average distance between p and q be $\leqslant x$. In this case, of course, one has not only $F_{pq}(0)=0$ for any two distinct lumps but even $F_{pp}(0)=0$. Another way would be to interpret $F_{pq}(0)$ as the probability that p and q overlap and then to proceed to the definition of distances by a chain method that will be discussed in a subsequent paragraph.

Especially remarkable is a probabilistic metric space studied by Schweizer and Sklar[6] which consists of random variables whose values are points in an n-dimensional Euclidean space E_n. The elements in this geometry may be thought of as cloud-like objects in E_n, which may not only overlap but completely interpenetrate. In the small, the behavior of lines is quite different from that of lines in a Euclidean space. In the large, the geometry asymptotically approaches that of Euclid.

Instead of elaborating on mathematical details of these interesting theories I here wish to discuss further steps suggested by the positivistic point of view. It certainly is in the spirit of Mach that these considerations comprise psychophysical as well as physical spaces, although from the excellent *Handbook of Mathematical Psychology* edited by Luce, Bush, and Galanter it appears that psychophysicists have not yet taken cognizance of, or at least not found use for, probabilistic geometry as a tool.

Unfortunately, the claim that the distance between two elements has (or probably has) a definite positive value does not in general seem to have a verifiable interpretation. If a unit has been chosen which is, say, approximately 100 times the size of the smallest discernible object, then there is no verifiable statistical correlate to the concept $F_{pq}(0.0001)$. But the more qualitative assertion that two elements p and q probably have *some* positive distance can be related to observations. If p and q are points, then $F_{pq}(0)=0$ may be interpreted as meaning that p and q are always found to be *distinct*.

Extending this idea one may look upon the inequality $0<F_{pq}(0)<1$ as meaning that there is a positive probability, namely $F_{pq}(0)$, that p and q cannot be distinguished, and the positive probability $1-F_{pq}(0)$ that they can be. Using some (more or less arbitrary) criterion one may speak of *probably distinguishable* and *probably indistinguishable* elements, e.g., according as $F_{pq}(0)\leqslant\frac{1}{2}$ or $>\frac{1}{2}$. More specifically, for any number i in the interval [0, 1], two elements p and q may be said to be *at least i-indistinguishable* if $F_{pq}(0)\geqslant i$. Statistically speaking, p and q are at least *i*-indistinguishable if they cannot be distinguished in at least the *i*-th part of the observations directed to these elements.

If $F_{pq}(0)=1$, then p and q are certainly indistinguishable as to their location in space. The case arises of course if p is a point and the symbol q is another designation of p. Interpenetrating elements, such as the cloud-like objects mentioned before, may well have the same location in space and yet be distinguishable in some extrageometric way. Further important examples arise in psychophysics. If an observer, Ob, at different times touches the same two spots, p and q, on the skin of a blindfolded person, Bl, with the two points of a compass, then it may well be that Bl experiences sometimes two sensations and sometimes only one. The relative frequency of the latter cases is the empirical correlate of $F_{pq}(0)$. For two spots that are very close to one another or very far apart, that number will of course be 1 or 0, respectively.

This probabilistic approach supplies us with the necessary complementation of Poincaré's ideas. The situation he symbolized by the formulae (1) may, along these lines, be described by

$$(2) \qquad F_{pq}(0)>0, \quad F_{qr}(0)>0, \quad F_{pr}(0)=0,$$

formulae that can hardly be called repugnant, let alone contradictory. All they express is the fact that a certain relation (which is neither the identity nor equality!) is not transitive. The relation in question is, to repeat it, the indistinguishability of two elements in at least some of the observations directed to those elements.

This relation is clearly defined in the experiment described above in which the reactions of the blindfolded Bl are superimposed on the geometry of the observer Ob – a geometry which, though itself strictly speaking only probabilistic, is considerably more subtle than that of Bl. The observer may draw a coordinate system on Bl's skin in ink in order to describe the location of the spots p, q, r, ... and, at any rate, it is Ob who decides whether the number of spots simultaneously touched on Bl's skin is 2 or 1. (Occasionally Bl reacts to a single stimulus with two sensations.) For Bl it would be, to say the least, much harder to construct a cutaneous geometry merely out of his own sensations. This problem leads back to the question of identifying elements. What for Bl would be the meaning of the relation of indistinguishability considering that in cases of indistinguishable sensations he has only one?

In cases where a relation of indistinguishability is clearly defined it can be made the basis of a probabilistic metric that is quite different from the one outlined before. For any two elements, p and q, one may determine the smallest integer n such that there exist n elements $p_1, p_2, \ldots, p_n$, which form a chain connecting p and q similar to the chains considered by Poincaré but with a probabilistic complementation of Poincaré's idea; that is to say, a chain such that $F_{p_k p_{k+1}}(0)>0$ for $k=0, 1, \ldots, n$ if we set $p=p_0$ and $q=p_{n+1}$. This integer n may be called the *probabilistic distance* between p and q. If $F_{pq}(0)>0$, then one may say that p and q have the distance 0. While a space with a probabilistic metric (i.e., with distribution functions F_{pq}) does not in general contain any finite chain joining two points, a space that deserves to be called a continuum does. G. Cantor defined connected sets by the property that any two points of such a set, for any number $d>0$, can be joined by a chain of points in which any two consecutive elements are less than d apart.

The preceding probabilistic definition of distance can be greatly refined. If p and q are two elements, then for each number i in the interval [0, 1] one may try to determine the distance between p and q by the probability standard i or, briefly, an *i-distance* $d_i(p, q)$ as the smallest integer n_i such that p and q can be joined by a chain of points satisfying the condition $F_{p_k p_{k+1}}(0)>i$ for $k=0, 1, \ldots, n$. Clearly, by increasing the probability standard one lengthens these chains; that is to say, if $i'>i$, then $d_{i'}(p, q)\geq d_i(p, q)$. The distance by the highest probability standard, i.e., the number $d_1(p, q)$ is the smallest number of points in a chain joining p and q such that consecutive elements are always indistinguishable – if there is such a chain; for it may well be impossible to uphold this highest standard of probability.

The i-distances so defined are integers that do not necessarily satisfy the triangle inequality. If $d_i(p, q)=2$ and $d_i(q, r)=3$, then $d_i(p, r)$ may be as much as 6. This actually is the case if the shortest chain joining p and r consists of the two points between p and q, the point q itself, and the three points between q and r. The numbers $d_i(p, q)+1$, $d_i(q, r)+1$, and $d(p, r)+1$, however, clearly do satisfy the triangle inequality.

Another example of distances defined by chains is contained in Schrödinger's colorimetric[7]. Schrödinger sets the distance between two colors, C and C', equal to 1 if C and C' are *barely distinguishable*, and equal to the integer n if there exists a sequence of elements, $C_0=C$, C_1, $C_2,\ldots$, C_{n-1}, $C_n=C'$ such that any two consecutive elements C_{i-1} and C_i are barely distinguishable while there does not exist any shorter chain of this kind. According to Schrödinger's assumptions about the color cone, each element C (represented by a vector in a 3-dimensional space) is the center of an ellipsoid whose interior points cannot be distinguished from C while the points on the surface of the ellipsoid are barely distinguishable from C. But such well-defined neighborhoods do not seem to exist in the color cone any more than on the human skin. One must define distance 1 in a probabilistic way and then can apply the idea of i-distances.

Perhaps the most difficult problem in microgeometry is, as I have already emphasized, the identification of individual elements of the space. In 1951, I suggested that, besides studying well-defined sets, one might have to develop a theory in which the element-set-relation is replaced by *the probability of an element belonging to a set.* In a note published in Paris[8], I called such an object, in contrast to an ordinary or rigid set, *ensemble flou* (=hazy set). In a slightly different terminology, this idea was recently expressed by Belman, Kalaba and Zadeh under the name of *fuzzy set.*[9] (These authors speak of the degree rather than the probability of an element belonging to a set.) Since, however, the concept of probabilistic sets still requires the identification of elements, I now believe that in geometry it ultimately will have to be combined with that of lumps, which can be more easily identified and distinguished than points. Moreover, lumps admit an intermediate stage between indistinguishability and apartness, namely that of overlapping. It is, of course, irrelevant whether the primitive (i.e., undefined) concepts of a theory are referred to as *points* and *probably indistinguishable*, or as *lumps* and *probably overlapping*. All that matters are the assumptions made about the primitive concepts. But the assumptions made in the two cases would not be quite identical and I believe that the ultimate solution of problems of microgeometry may well lie in *a probabilistic theory of hazy lumps*. The essential feature of this theory would be that lumps would not be point sets; nor would they reflect circumscribed figures such as ellipsoids. They would rather be in mutual probabilistic relations of overlapping and apartness, from which a metric would have to be developed.

Since, as I have already intimated, a positivistic geometry cannot start with anything ultimate it must start in the middle, which means that it must lend itself to extensions in two directions – to more and to less refined theories.

(1) There should be a formalism for the refinement of the geometry corresponding to the potential refinements of instruments and methods of observation, typified by the transitions from observations with the naked eye to the use of magnifying glasses, of microscopes, of electro-microscopes, ... A refinement of a geometry resolves some of its indeterminacies in the small. The elements P, Q,... of the geometry – points or lumps – are replaced by finite sets of elements, $p, p', p'', \ldots; q, q', q'', \ldots; \ldots$ These sets need not be disjoint. How the distance between P and Q should be related to the distances between the p, p', p'',... and the q, q', q'',... has not yet been fully clarified.

(2) There should be a formalism of superimposing a less refined on a finer geometry corresponding to the blindfolded person's superimposing his subjective geometry of skin sensations on the observer's geometry of that surface (i.e., the observer's idea of points, distances, betweenness, and the like). Sets of points or hazy sets of points of the observer are combined into single sensation points of the blindfolded man. These sets are not in general equivalence classes since they need not be disjoint and, as has been remarked, indistinguishability on various probabilistic levels is not a transitive relation.

As far as the physical space is concerned, the outlook for the application of a positivistic geometry to atomic and particle physics is not at present very bright. For those physicists who express interest in the idea wish – understandably! – that mathematicians develop a large number of different geometries from which they (the physicists) can choose the one most suitable to their purposes. But it seems to be difficult enough to develop one geometry of such an entirely new type. The statements of macrogeometry should result from assumptions about space in the small through averaging processes somewhat like macrocosmic gas phenomena result from assumptions about molecules in kinetic gas theory. Therefore, one of the guiding principles in formulating those assumptions about space in the small should be comparable to the correspondence principle in Bohr's atomic theory and guarantee that, in the large, the geometry asymptotically approach that of Riemann. Although such a development has hardly begun one may perhaps venture a certain hope that one day proper assumptions about space (and time!) in the small may geometrize microphysics somewhat like Riemann's geometry in the large has geometrized gravitation in the general theory of relativity.[10]

II. THE CLARIFICATION OF FUNCTIONAL CONNECTIONS

In dealing with the application of a branch of mathematics to science the methodologist has to clarify three points: the mathematical concepts to be applied; the nature of the scientific material; and the method of application. Mach's proposal to replace causal explanations by functional connections, therefore, raises three questions: (1) what are functions? (2) exactly what is it that functions are supposed to connect? (3) how is that connection performed or achieved? Traditionally, these questions have been answered as follows.

(1′) A function is a *law* or a *rule* associating a number, denoted by $f(x)$, to each number x belonging to a certain class of numbers, or associating a number $F(x_1, x_2, \ldots, x_n)$ to each n-tuple $(x_1, x_2, \ldots, x_n)$ of numbers of a certain kind. The symbols x and $x_1, x_2, \ldots, x_n$ in the preceding definitions are referred to as *variables*; e.g., the function associating x^2 with x and the function associating $x \cdot y$ with (x, y) are said to be functions *of one variable*, more specifically *of* x, and *of the two variables* x *and* y, respectively. They often are called 'the functions x^2 and xy' or denoted by the so-called equations $y = x^2$ and $z = x \cdot y$.

(2′) Functions are applied to *variables* or *variable quantities* (assumed to be known without explicit definitions except that in some cases units are mentioned), which include the most important objects of traditional geometry and physics such as s and a (length of a side and area of squares) and v, p, t (volume, pressure, and temperature of ideal gases).

(3′) The application is expressed in *formulae* such as $a = s^2$, or (in proper units) $t = v \cdot p$. These formulae are interpreted as summarizing statements of the form: if $s = 3$, then $a = 3^2$. The variables s, v, p in the preceding formulae are said to be *independent*; a and t, *dependent*.

This presentation of the situation calls for several critical remarks and for a complete reformulation.

(1″) Laws and rules are metamathematical rather than mathematical concepts. Mathematicians do not add and multiply laws while they do add and multiply functions. Some mathematicians, therefore, replace (1′)

by the statement that a function is defined if, with each element x of a certain set of numbers, exactly one number, denoted by $f(x)$, is somehow associated. But, strictly speaking, this is a definition of the *definition* of functions rather than a definition of *functions*. The objects, moreover, of mathematical theorems are not acts of somehow associating or pairing numbers to numbers, but rather the results of such acts, that is, sets of pairs of numbers. This (extensional) point of view in mathematics can be traced, if not further, certainly to Osgood, who in 1906 mentioned it in a footnote on p. 1 of his *Lehrbuch der Funktionentheorie* without, however, utilizing the idea. It was widely adopted in the 1930's. Ever since, mathematicians have considered a binary relation as a set of ordered pairs of elements, and a function as a set of ordered pairs of numbers not including two pairs with equal first and unequal second members. If f is such a set, then the set of all numbers that occur as the first members in a pair belonging to f is called the *domain* of f – briefly, dom f. Since, if x belongs to dom f, there is only one pair in f whose first member is x, the second member of that pair, being determined by f and x, may be unequivocally called $f(x)$ or the value of f for x.

In presence of the clear and simple symbol f for the function, it seems to be absurd to refer to it rather as *the function* $f(x)$ or *the function* $y=f(x)$ or, for that matter, to the function log as the function log$x$. The only excuse is the fact that in case of the function referred to as x^2 or $y=x^2$ a clear and simple symbol for the function is lacking – a lacuna in the symbolism of analysis that can be traced to the 17th-century fathers of this branch of mathematics, who left the most important of functions, the *identity function* (i.e., the set of the pairs (x, x) for all x) as well as the constant functions, anonymous. But, as I have repeatedly emphasized[11], this can be helped by adopting a permanent symbol for the identity function, say j, which is actually being done in an increasing number of recently published textbooks and, I venture to predict, will be a very common practice one or two generations hence. For, in possession of such a symbol, one can speak of the function j^2 just as one occasionally speaks of the function f and might speak of the functions log and cos. One then can dispense with the equivocal symbol x^2 for the function j^2 (reserving it for the value that j^2 assumes for x) and can eliminate those so-called equations which actually are not treated as such, since the sum of the functions misnamed $y=x^2$ and $y=x$ is the function $y=x^2+x$ and not $2y=x^2+x$. Moreover, in this way, the function concept is prepared for the role of a connective which, as will be seen in (3″), it plays in science.

In defining j^2 as the set of the pairs (x, x^2) for all numbers x one uses x as a variable, more specifically as a *number variable* in the sense first clearly defined by Weierstrass in his lectures in the 1870's, that is, as a symbol that stands for any element of a certain set of numbers. If in the definition of j^2 the variable is replaced by the designation of an element of the set of all numbers, this replacement results in the designation of an element of j^2 such as $(3, 3^2)$. At the same time one realizes that it would be absurd to call j^2 (or, for that matter, log or f) a function *of* x or *of* a variable or, in fact, of *anything*.

(2″) The most important task of the clarification of applied analysis is to overcome the century old misidentification of what are called variables, or variable quantities in science, for example, s, a, v, p, t, with variables used in mathematics and logic, as first defined by Weierstrass. As I have mentioned above, traditionally it is assumed that the so-called variable quantities are known and that the symbols for them such as v, p, t are understood without either definitions or postulates concerning them. The closest to a definition is a rule stipulating how one should operate with them in pairs or triples as they appear in formulae such as $t=v\cdot p$; for example, the rule that if v assumes the value 3, and p assumes the value 4, then t assumes the value $3\cdot 4$. But what are the individual variable quantities v, p, t? Being denoted by separate symbols they ought to be separately defined. Yet all that is said about them individually is the obscure hint that they are *variable numbers* assuming different values in different cases.

In what misunderstandings these confused indications have resulted is illustrated by the following remark in the first section of Tarski's *Introduction to Logic*: "... it is said that the symbols 'x', 'y',... also denote so-called 'variable numbers' or rather 'variable quantities'... The properties of such a number would change from case to case: that is to say, the number would sometimes be positive, sometimes negative, and sometimes equal to 0. But entities of such a kind we do not find in our world at all; their existence would contradict fundamental laws of thought." While of course a number cannot be sometimes 3 and sometimes 4, Tarski seems to be quite unaware of the fact that the words variable quantity in the literature do not really refer to numbers or symbols for numbers but rather are references – of course obscure and confused references – to a totally different category of objects, including v, p, and t, each of which indeed can sometimes have the value 3 and sometimes the value 4. I have revived Newton's term *fluents* for the entities of this category[12].

Preliminary separate definitions of the fluents v, p, and t can be formulated as follows. Given units, we pair with each sample γ of an ideal gas three numbers, $v(\gamma)$, $p(\gamma)$, and $t(\gamma)$. The gas volume v is the result of the first of these pairings, that is, the class of all pairs $(\gamma, v(\gamma))$ for all samples γ of an ideal gas. Similarly, I define the fluents p and t. This definition of fluents resembles that of functions. But there is a fundamental difference. dom j^2 consists of numbers, which makes j^2 a purely mathematical concept; dom$v$ consists of gas samples, which makes the concept v extramathematical. J. Mehlberg[13] has described the difference between fluents and functions from the point of view of the theory of types. Another difference lies in the fact that only functions (and not fluents) can serve as connectives and, therefore, are used throughout science, while fluents belong to specific branches of science and cannot connect either fluents or functions. We speak of the logarithm of the temperature, the logarithm of the mass, and so on; but we cannot, as I have pointed out before, speak of the temperature of the logarithm or of the mass of the temperature.

Some algebraists seem to take the position that since both functions and fluents, as herein defined, are mappings of a domain into a set of numbers, both deserve the name of functions in the wider sense now often adopted; and that nothing new has been achieved by the definition of fluents herein propounded. What those mathematicians seem to overlook (apart from the differences in type between fluents and functions) is the very essence of the clarification: that its main point is not the introduction of a new term for variable quantities (although even this in itself might be useful in forestalling that fluents be mistaken for variables); but that its essense is a manageable definition of a category of extremely important scientific objects (namely, of fluents), which previously were ill-defined or altogether lacking any definition. The fact that a gap in the ontology of applied mathematics has been filled may be unimportant for algebraists but, as will be seen in (3″), it certainly brings about a complete reinterpretation of century old formulae such as $a=s^2$ and $t=v\cdot p$.

(3″) Just as a number connects one object with another object, or one number with another number, e.g., as 12 connects, ratiowise, years with months, feet with inches, 36 with 3, and so on, a function connects fluents with other fluents, or functions with other functions; e.g., the product function assuming the value $x \cdot y$ for (x, y) connects t with v and p in the realm of ideal gases, the area with length and width in the realm of rectangles, the sine with the cosine and the tangent, and so on. If the product function is denoted by P, then one may describe the gas law by

(3) $t = v \cdot p$, or $t = P(v, p)$.

Formulae (3), which connect the fluents v, p, and t, are, in view of the definitions of the latter, summaries or abbreviations of the following formulae, which connect numbers, more specifically, values of those three fluents:

(3_γ) $t(\gamma) = v(\gamma) \cdot p(\gamma)$ or $t(\gamma) = P(v(\gamma), p(\gamma))$,

for each element γ of the domain of v, p, and t. Here γ is a variable, i.e., a symbol that stands for any element of a certain class – the class of all samples of ideal gases. If in (3_γ) the variable is replaced by the designation of the contents at a certain instant of a container at a certain place, then this replacement results in a statement about that particular gas sample. There is a perfect parallelism between (3_γ) and the following formulae connecting numbers that are values of functions

(4_x) $\sin x = \cos x \cdot \tan x$ or $\sin x = P(\cos x, \tan x)$,

for each number x belonging to dom tan. The functions cos, tan, and sin correspond to the fluents v, p, and t; the number variable in the mathematical formula is the counterpart of the sample variable in the physical law. The century-old methodological misinterpretation of physical formulae was based on the invalid supposition of a parallelism between the variable x in (4_x) and the fluents v and p in (3). (That the latter were also called variables seems to have been partly a reason, partly a consequence, of the way the formulae were interpreted.) All that has to be done in order to bring out a complete parallelism between the mathematical and the physical formulae is to formulate an abbreviation of (4_x) that connects the functions themselves, summarizes the statements about their values, and thus corresponds to (3), that is,

(4) $\sin = \cos \cdot \tan$ or $\sin = P(\cos, \tan)$.

That in the 17th century mathematicians wrote only (4_x) and not (4) was one of the corollaries of their failure, mentioned in (1″), to introduce symbols such as j and j^2 for the most important functions. Hence they could not write

$$j^3 = j^2 \cdot j, \quad \text{or} \quad j^3 = P(j^2, j),$$

but had to write $x^3 = x^2 \cdot x$. So, for the sake of uniformity, they did not go beyond (4_x). Thus, while physicists and applied mathematicians never write (3_y), the diehard adherents of the 17th-century mathematical symbolism even to this day eschew formulae such as (4).

From what has been said it follows that one may well speak of a function of one or of several fluents. But it is clear that in so doing one refers to fluents and not to functions; e.g., the product of v and p is the fluent t. Hence this locution in no way contradicts the remark in (1″) that a function is not a function of anything. There exists in this respect a parallelism between functions and numbers. Just as we speak of the logarithm of the temperature we speak of 12 apples. But only young children are confined to the study of numbers of apples or objects of any kind. The adolescent learns to abstract the idea of number and to use numbers as connectives. The arithmetician never mentions numbers of extramathematical objects at all (though he may apply the number concept, e.g., in order to count the number of prime numbers below a certain number). Speaking of functions only in connection with specific fluents or even in terms of fluent variables (i.e., symbols that stand for any fluent of a certain kind) is a remnant of the childhood of analysis. The analyst never makes any reference to extramathematical fluents though he may use a function in order to connect other functions. (Incidentally, if a formula expresses a fluent as a function of one or several other fluents, then these other fluents correspond to what are traditionally called independent variables, the first, to the so-called dependent variable. All this, of course, is relative to the particular formula since there may well exist another formula expressing the second fluent as a function of the first.)

A few further remarks should be devoted here to that part of the theory which is closest to positivistic ideas, that is, the concept of fluents. The definition of some fluents can be modified by variations of their domains. For example, one may associate the values of v, p, and t with states of gases rather than with gas samples. (This modification has nothing to do with the fact that gas samples change in time since, as goes without saying, the term sample refers to what sometimes is called time slice of a sample, i.e., a sample at some instant.) With regard to the modified definition of the gas fluents, naturally the question arises as to how the state of a gas can be described and ascertained. The traditional answer to this question maintains that the state of a gas is determined by the values of v, p, and for that gas. This solution of the problem is not necessarily circular as it may seem to be at a first glance. But it leads to a complete arithmetization of the material and altogether detaches it from reality. For the definition amounts to an identification of the fluents with functions in the strict mathematical sense of associations of numbers with triples of numbers. More specifically, v, p, and t are in this way defined as what are called the first, the second, and the third *selector* functions, whose values for each triple of numbers are the first, the second, and the third member of the triple, respectively. A very similar procedure is followed in pure analytic geometry, where points are introduced as triples of numbers and, say, the abscissa is defined (if at all!) as the first selector function. On this foundation (which may well be said to be neo-Pythagorean since Pythagoras taught that everything is number) one can of course develop extensive algebraic and analytic theories; but in order to apply the results to reality one must leave this panarithmetical universe and must formulate procedures connecting the coordinates with observable entities; that is to say, one again faces the problem of defining fluents. In a similar way, the quantum mechanical definitions of observables as operators must be supplemented, and for the sake of applications are being supplemented, by the formulation of procedures connecting the operators or probabilities computed by means of the operators with *observable* observables.

The most positivistic domains which, as I have repeatedly emphasized[14], can and, in some cases, must be considered, consist of acts of observation, especially, acts of meter readings. The value of a fluent for such an act α is the number read as the result of α. If the domains of v, p, and t consist of volumetric acts β, and acts μ and θ of manometer and thermometer readings, then it is clear that (3) cannot be interpreted as

$$t(\theta) = v(\beta)\cdot p(\mu) \text{ for any acts } \beta, \mu, \theta.$$

The validity of this formula presupposes a correlation of the domains of v,

p, and t, namely, that β, μ, and θ be acts simultaneously directed to the same gas container.

In general, that is, in terms of fluent variables (which, I repeat, are symbols that stand for any element of a certain class of fluents), the simple functional connection of fluents can be described as follows. If u and w are fluents, if f is a function, and if Π is a set of ordered pairs (α, β) such that α belongs to dom u and β belongs to dom w, then

$$w = f(u) \text{ rel. } \Pi,$$

if and only if $w(\beta) = f(u(x))$ for any pair (α, β) belonging to Π. Since f is a set of pairs of numbers, this formula is equivalent to the statement that $(u(\alpha), w(\beta))$ belongs to f.

Until recently I have found in discussions that explicit references to acts of observation in the formulation of the laws of classical physics have been unacceptable even to applied mathematicians. It certainly was an important step toward positivism when in 1966 Wigner and collaborators of his[15] included just such explicit references to acts of observation in laws of atomic physics. These are among the formulations alluded to in part 1 of this paper, which express probabilistic predictions of the results of future acts of observation on the basis of the results of such acts performed in the past. The clarification of the statements of classical physics, which has been briefly summarized in this paper, pointed, as far as macrocosmic laws are concerned, in this direction which certainly is in line with thoughts of Ernst Mach.

Illinois Institute of Technology

REFERENCES

[1] *Rice Institute Pamphlets*, **27**. No. 1 (1940) 80–107.
[2] K. Menger, 'Statistical Metrics', *Proc. Nat. Acad. Sci. USA* **28** (1942), 535–537; 'Probabilistic Geometry', *ibid.* **37** (1951), 226–229, and 'Géométrie Générale', *Mem. Sci. Math. Paris*, **124** (1954).
[3] A. Wald, 'On a Statistical Generalization of Metric Spaces', *Proc. Nat. Acad. Sci. U.S.A.*, **29** (1943), 196–197.
[4] B. Schweizer and A. Sklar, 'Espaces métriques aléatoires', *Comptes Rendus Acad. Sci. Paris*, **247** (1958), 2092–2094; and 'Statistical Metric Spaces', *Pacif. J. Math.* **10** (1960), 313–334.
[5] A. N. Šerstnev, 'On the Concept of a Stochastic Normalized Space', *Dokl. Akad. Nauk SSSR* **149** (1963), 380–283.

[5a] For a brief summary, see B. Schweizer, 'Probabilistic Metric Spaces – The First 25 Years', *The New York Statistician* **19** (1967), 3–6.
[6] B. Schweizer and A. Sklar, 'Statistical Metric Spaces Arising from Sets of Random Variables in Euclidean *n*-Space', *Teoria Veroyatnostey* **7** (1962), 456–465.
[7] E. Schrödinger, *Ann. Phys.* **63** (1920), 397–456.
[8] K. Menger, 'Ensembles flous et fonctions aléatoires', *Comptes Rendus Acad. Sci. Paris* **232** (1951), 2001–2003.
[9] R. Bellman, R. Kalaba, and L. Zadeh, 'Abstraction and Pattern Classification', *J. Math. Anal. Appl.* **13** (1966), 1–7.
[10] K. Menger, 'The Theory of Relativity and Geometry', in *Albert Einstein: Philosopher-Scientist*, Evanston, Ill., 1949, pp. 472–474.
[11] E.g., in 'An Axiomatic Theory of Functions and Fluents' in *The Axiomatic Method* (ed. by Henkin *et al.*), Amsterdam 1959, pp. 454–473; 'Mensuration and other Mathematical Connections of Observable Material' in *Measurement, Definitions and Theories* (ed. by Churchman and Ratoosh), New York, 1959, pp. 97–128.
[12] Cf. the papers quotes in [11] and 'Variables, Constants, Fluents' in *Current Issues in the Philosophy of Science* (ed. by Feigl and Maxwell), New York, 1961 pp. 304–316; 'A Counterpart of Occam's Razor in Pure and Applied Mathematics', *Synthese* **12** (1960), 415–428, and **13** (1961), 331–349.
[13] J. Mehlberg, 'A Classification of Mathematical Concepts', *Synthese* **14** (1962).
[14] Cf. especially the second paper quoted in [11] and 'Rejoinder to Adams' in *Current Issues in the Philosophy of Science*, quoted in [12], p. 318.
[15] Houtappel, Van Dam, and Wigner, 'The Conceptual Basis and Use of the Geometric Invariance Principles', *Rev. Mod. Phys.* **37** (1965), especially pp. 598–600.
Added in proof: For explicit references to procedures, see also H. Ekstein, 'Presymmetry', *Phys. Rev.* **153** (1967), 1397–1402, and 'Presymmetry II', *Phys. Rev.*, Aug. 25, 1969.

ROBERT S. COHEN

ERNST MACH: PHYSICS, PERCEPTION AND THE PHILOSOPHY OF SCIENCE*

I. PREFACE

Though a working physicist, Ernst Mach confronted questions not only of physics but also of psychology, biology, scientific methodology, the history of ideas and of technology and general culture, and metaphysics. No doubt he was unusual, but still it was as a physicist who sought to understand physics in unusual depth that he found he had to understand both the logic and the history of concepts; and his understanding of the relation of physics to reality led him to study observation, measurement, and psycho-physics. It would be useful to trace Mach's thought generally from physics. However, others have undertaken this task in the symposium and elsewhere; in this essay[1], I shall discuss his philosophy of science by a more direct route.

II. PHENOMENALISM

Mach is best known as a phenomenalist, and his conception of science is often taken as a simple cliché: the world consists of sensations, for the scientist and for the common man. But it is not a simple matter to understand this assertion; and it is our principal task here to do so with sympathy and with scepticism. Some of my readings of Mach are fairly conventional, others seem unorthodox. The crux of any unorthodoxy, usually is my sceptical sympathy, especially my sympathy with the hardly noticed, though conspicuous aspect of Mach's philosophy, that is the impersonal serene philosophy which goes beyond subjectivism and objectivism. Before coming to this, however, we set forth Mach's intellectual motivation; so as to see him for what he really wanted as a philosophical interpreter of science. All the rest simply developed out of this motivation.

He was eager to criticize what is illusory in science and common sense, and by so doing, better to delineate what is genuinely known. Astonished as a youth by his somewhat accidental discovery of Kant's critical philosophy, as presented in the *Prolegomena,* he later studied Hume and Darwin, and thus learned to approach his beloved physics with a wary and sceptical eye. Ancient philosophical problems soon re-appeared for him within the working habits of science such that the meanings of the elementary aspects of his professional practices and teachings had to be questioned. What is an observation? an experiment? a measurement? What is mass? force? space and time? What is a law of nature? a theory? an explanation? To these questions, he brought his rigorous skills in physical experimentation and in mathematical analysis, but in addition he taught himself, from his earliest scholarly days, to examine the logical and empirical characteristics of concepts through elucidation of the time sequence of their various formulations and modifications.

Other able investigators, incidentally, have followed the same paths of historical inquiry, some with notably different conclusions about the nature of science – such as Émile Meyerson – and others in agreement – such as Pierre Duhem. The history of scientific conceptions is not easily revealing, and to its study the investigator brings his own purposes and a conceptual apparatus which, although it may be modified with use, sets – regrettably or otherwise – something of his overall views and pertinent conclusions.

What, then, may fairly be claimed as Mach's views? In response to sceptics, he wished science to assert only that for which there can be evidence. Like so many empiricists, but perhaps more doggedly, he believed: not speculation, not insight, but rather bodily sensations are the source of evidence, in whatever field we may inquire. Sensations cannot reveal what is closed to sensation, nor can we infer knowledge of what is behind sense perception. The spatial metaphor 'behind sensation' suggests a meaning but it cannot be shown (or sensed) as applicable. Whereas this metaphor of 'behind' is out, another, however, is in: 'shared'; sensed data are the common shared mode of evidence, for every observer and for each science. Hence, solipsism is ruled out *a priori.* So is the watertight compartmentalization of science. Indeed unification of the sciences is logically and epistemologically at least feasible, if not more. One might ask, how can Mach reject one metaphor and embrace another? Rather than confront this difficulty of the metaphor 'share' in the statement, 'Science is shared by all observers', Mach pressed it further himself by asking (with

Hume) whether 'observer' was not a metaphor as well. The self, or ego, is a puzzling notion, at least by no means evident, by no means shown by sensation. Indeed the apparently private and ego-referring character of sensations, which, incidentally, is what led Mach's predecessors inexorably to solipsism, sometimes against their protests, *is* misleading since scientific knowledge is public in reference, in usefulness, and in testability; and so Mach replaces 'sensation' by the neutral word 'element' which commits him to assertions neither about objects (which his critics would consider to be shown by sensations but to exist independently) nor about ego-centric worlds (which would consist of sensations and would risk solipsism). To introduce the notion of neutral 'elements' at first doubtless seems only a ploy, but it is Mach's primary entry to his philosophical achievement. For him, science is an attempt to provide descriptions of the 'elements' in a language of functional dependence, economically, comprehensively, simply. Such descriptive language, aside from giving esthetic satisfaction, often yields predictions, and with them some control over nature – or should we remind Mach to say, control over the elements? In this respect, scientific ideas are tools for (Darwinian) survival of the individual and of the human species. By sharing knowledge in the human community, and through other communities, the individual transcends the limits of egoism, for Mach envisioned the world as a whole, and man's place as passing but yet integrated.

One may wonder whether it is not the case that one metaphor – 'ego' – was merely displaced by another – 'whole' – with little or no anti-metaphysical profit. Be it so or not, one point seems clearly to be correct: for Mach, the profit has to be anti-metaphysical.

As we remember, Mach's principal motive was to delineate knowledge from false claims to knowledge, and he took it for granted that knowledge was sensational; and hence that metaphysics was speculative, with no true claim to be considered as knowledge. Consequently all its claims for certitude cannot be considered as anything less than dogma. One might question, why does Mach claim for metaphysics the status of certitude which enables him to unmask it as dogma? Before attempting to answer this, we must note that the great metaphysicians, from Descartes to Kant, did claim certitude for metaphysics – and the great anti-metaphysicians, from Bacon to Duhem, shared this unmasking. Meyerson, as I have indicated elsewhere[2], is in a class by himself. Historically, the answer seems to be that metaphysics, along with science, was, in the West, a reaction to scepticism. When metaphysicians speculate, and present a frankly tentative as well as speculative view, as with Boscović, they have aroused neither the same hostility nor, regrettably, the same interest either.[3]

Mach, then, wished to overcome all metaphysical dogmas which were alleged to provide, in their various ways, nothing short of certainties. These certainties were, however, untested in experience; in fact, Mach thought he could unmask them with a new certainty. The traditional anti-metaphysicians stressed the uncertainty of the conclusions of metaphysics, of the doctrines themselves; and this uncertainty derived, of course, from the speculative method. Contrary to this, it was claimed, the observational method led to assured conclusions. Mach deviates from this. He economizes: instead of going to conclusions via method, he stresses method itself. The conclusions of science, hence, will indeed be tentative; but its method is not subjective. To call scientific method 'objective' will, as Mach sees it, run the risk of rendering the 'elements' as *the given*, and scientific conclusions final and absolute. The method of science, to Mach, is neither subjective, as that of metaphysics, nor objective as the old empiricism claimed, but impersonal. (Mach's followers in Vienna used intersubjectivity for impersonality, which is unwittingly either Kantian or diluted Machism; still worse, it may be old empiricism with its solipsistic pitfalls.[4])

For Mach, then, not knowledge attained, but the method of attaining it, could be certified. Indeed, knowledge is always provisional, in Mach's understanding, but nevertheless it is constructed upon an indubitable and elementary base – indubitable, that is, in the sense of being unavoidable, not in any sense of being unalterable. All conceptions which fail to relate ideas to that base are thereby suspect; their meaning is ultimately unclear, for they might be conjectures about phenomena which as yet are misapprehended, or whose apprehensions are mis-formulated, or they might be pseudo-intelligible assertions which, upon scrutiny, lose meaning, and thus lose all possible experiential reference. And such pseudo-intelligibility is the mark of metaphysics. In his historical researches, Mach traced metaphysical and empirical aspects of scientific ideas, but his philosophical analysis of science was dependent less upon his destructive critique of metaphysics than upon his constructive directives for science.

The latter, in turn, were extracted from his own scientific efforts and his historical studies; and then generalized.

To Mach, then, metaphysics is an illusion of knowledge; a deception of the human spirit. It deceives not only in what it claims, but also in a peculiar side-effect: it has typically injected dualism of mind and matter into the natural world; of nature and super-nature into the cosmos. To cut through such doctrines, and to overcome any absolute dualism (not to mention pluralism) whatever, monism is not sufficient, since any monism may be ephemeral; moreover, it will be an adequate or inadequate alternative to dualism, not cut through all possible dualisms. To do that requires a monistic criterion which is to be used in any case of philosophical criticism of any dualism or pluralism of any sort. Consonant with the trend of modern European thought, Mach saw this as an epistemological problem. How, to take as a mere example the crucial dualism of mind and matter, may we construct a model of reality so that it consists neither of one nor of the other exclusively, and yet also such that it is not composed of mutually exclusive factors? Mach's plan was simple to a fault, and his argument too. The evidence for the material – things or events – is displayed before us by observation, and the evidence for the mental likewise. Thus we know of the mental and also of the material from within our own bodies, and from our bodies: in both cases with an immediacy which requires no inference and which has no publicity. But, setting privacy apart from discussion for the moment, observational evidence for material events is identical with that for mental events of other persons: we might talk of the sensation language of colors, sounds, smells; or of the mathematical language of spaces, times, shapes; or of a thing language of bodies, pressures, temperatures. Mach uses these, often interchangeably, but his primary procedure is to re-define – presumably deliberately – any scientifically conceived entities by a mixture of vocabularies – presumably provided as alternative. From this viewpoint, it is unscientific and unreasonable to talk about things which interact with the mind, and by this interaction produce sensations. The colors, sounds, spaces and times are themselves the typical elements of reality with which science can deal, and it is our scientific purpose to investigate them. In Mach's well-known phrases, "bodies do not produce sensations, but complexes of elements make up bodies ... all bodies are but thought-symbols for complexes of elements." And in another passage, "sensations are not signs of things; but, on the contrary, a thing is a thought-symbol for a compound-sensation of relative fixedness."[5]

So Fichte's turn against Kantian things-in-themselves – itself a positivist critique of a metaphysical part of *the* critical philosophy – was used by Mach too, but just as harshly against the Fichtean 'self', the metaphysical element still present in Fichte's positivism.[6]

Pure experience is Mach's epistemic base, purified, as it were, of any reference to a source, as well as to a recipient, or, *a fortiori*, to any causal interaction. Nevertheless, Mach's formulations are just as often mere cautions as they are nakedly positivist. In one typical exposition, he describes colors, sounds, spaces, times, as being only "provisionally the ultimate elements".[7] He seems clearly to wish to distinguish between a phenomenalist confirmation base (which is also the phenomenological presentation of the 'world') and any dogmatic certainty about the fundamental nature of the world. We are not, of course, assured that his methodology will permit even such minor irresolution about ontology, but his intent was clear.

Though clear, his intent was complicated by his empirical attitude to knowledge as a phenomenon. Distinct from his purely rational reconstruction of scientific concepts, Mach's appraisal of their very nature was biological and historical: he considered concepts to be properties of the concept-making animal. Man, as animal, may be a thing, and things, as we have quoted above, are "thought-symbols for (elements) of relative fixedness"; or, man may be the measure of all elements, thereby being their dual counterpart. Let us not press this just now. For, one way or another, it is in this very consideration that Mach was a pioneer, transforming epistemology from philosophy to science. He was a worker in psycho-physics, and a close student of all studies of interactions between biology and physics. Evolutionism impressed him profoundly, and likewise researches into ancient and prehistoric technology. With a cultural appreciation of science and technology as instrumental, and the then current biological conjecture that knowledge is one form of adaptation to the environment, Mach was less likely to enlarge his positivist epistemology to cosmic scope than were the old-style empiricists, whether Baconian or Humean. His positivism was a scientific theory of how and what scientists do when they are engaged in the practical activity of exploring, reflecting, calculating, checking, refuting, knowing. Seen in this new bio-historical context, Mach's theory of science is a new and better version of Bacon's counsel of modesty, and warning against *hubris*. In fact, he seemed to conclude, science is not an attempt to understand the world as it is but only to describe the world as we experience it; and epistemology, to be scientific, must likewise be not an attempt to understand the phenomenon of science but only a description of it. Generally, then, understanding is not an attempt to master the universe but only to help us in our cosmically local way through life. To put it differently, mastery is not cosmic as in Bacon but localized. Our human perspective on the world provides the scope of science; there is no cognitive alternative to science, just because there is no cosmic perspective to provide a greater scope. And so Mach's phenomenalism is a hard-headed positivistic attempt to find an alternative sort of objectivity within a species-centered outlook upon phenomena.

Species-centrality is more than subjectivity, and more even than intersubjectivity; but it is less than finality or objectivity in the sense of final knowledge of what there is – as Bacon, Descartes, and even Hegel would have it. I have called it 'impersonalism'. In the end, Mach's new reinterpretation of laws and theories in the light of his new idea of impersonality is that of compressed descriptions of experience. This reinterpretation, one may object, gives no adequate account of the referential meaning of Darwin's evolutionary hypothesis, nor of the facts and theoretical interpretations of human history. If so, clearly, the Machian epistemology would be blocked by this self-referential impasse. Indeed, we have already met this problem before, when asking whether Mach is not a dualist after all. It is still too early to confront this point; before we do so, we must first see how his phenomenalism meets a number of questions within work-a-day science.

Among the many elements (in Mach's sense) within the vast flux of observables, rational man will seek correlations, and his goal is the "fixing of the flux of the elements, whether mediately or immediately".[8] In such a view, all is flux, but we are not justified to say that all is *logos* or process. For, relative fixity which we find among elements does not point beyond itself. Nor do our linguistic and mathematical symbols for such stable correlations refer to more than the observed correlations. So if appearances are connected, we must carefully distinguish the fact of connection from any hypothesis about substance and form, entity and property, matter and its governing laws – all of which may be imagined to explain the observed connections. Hume on causation hints at the pattern Mach follows, at least as a start, in his understanding of all connective correlations among sensed elements. But though this is correct, Mach goes beyond Hume's sceptical abandonment of any reasoned basis for science. The connections among observed elements do indeed exist, for Mach, since we can show them; but "what we represent to ourselves behind the appearances exist *only* in our understanding."[9] The question presses itself, what kind of existence is that? and momentarily we suspect that Mach may wish wholly to separate ordinary perceptions from the elements, from the *pure* sensations and *mere* observations. And, indeed, he does so, but only instrumentally, and not as his explanation of the existence of (at least seeming) scientific explanation. What we "represent ... behind the appearances" he goes on to say, has "only the value of a *memoria technica* or formula whose form, because it is arbitrary and irrelevant, varies very easily with the standpoint of our culture."[10] He uses the word 'because', but only metaphorically. Let us proceed slowly.

Any formula, he says, and with it any so-called law of nature – however thoroughly it may have been tested, confirmed, and applied – is of no greater value than that of all the individual 'facts' (his word) which it describes. Formulas and laws obviously are convenient; and they are convenient because they are economical. That is as far as Mach will go: as he sees the history of science, no explanation will show why it is that the elements can so conveniently be organized, nor, to put the matter the other way, why laws and formulas work: Mach, the scientific epistemologist, *merely* records the fact. This is a difficulty all instrumentalists have to face: they have no room in their philosophy for explanation in general, nor for explanation of scientific success in particular. Mach at least knows this fact. Cultural relativism plus formalism cannot explain the phenomenon of science.

To this truncated view of Mach as well as all instrumentalists and all relativists, the further query may be posed: how shall he incorporate the use of hypotheses, so common among scientific theorists? Must we reconstruct science so as to eliminate whatever is conjectural? forbid submicroscopic entities altogether? deny any role to entities which transcend the phenomenal given? If so, then we cease describing and start stipulating. But Mach, whose knowledge of physical theory was deep, could scarcely have ignored the place of hypotheses in science, nor stipulated for science, especially not driven physics back to inductive or abstractive techniques alone. And in fact, in a manner somewhat reminiscent of Bellarmine three centuries earlier, he set no restriction upon free play of mathematical imagination.[11] Whatever may be useful, may be used; but the meaning of imagined entities and theoretically conceived ideas needs careful investigation and test. In a famous passage[12], Mach wrote

If the hypotheses are so chosen that their subject (*Gegenstand*) can never appeal to the senses and therefore also can never be tested, as is the case with the mechanical molecular theory, the investigator has done more than science, whose aim is facts, requires of him – and this work of supererogation is an evil.

And so theory is not always forbidden, although – remember his anti-metaphysics – it *may* be forbidden. It must be testable. When? Mach does not say, and in some cases the gap between the act of theoretical imagination and the act of experimental test will be of long duration. But if any theory can be so analyzed that, in Mach's words, it "can never be tested", then clearly it is not relevant, nor even meaningful, in our world of experience. Or, we may understand Mach to say, a theory for which all presently conceivable experience is irrelevant is at least not yet a part of science. We can, of course, make mistakes in our analysis, and we may demand tests too hastily: the chains of empirical relevance may be subtle indeed. For example, Mach offered powerful arguments, from the positivist stock, against the atomic-molecular theory, but, while as we know he erred in rejecting the theory for so long, he would have been persuaded by any direct or indirect evidence of the theory's superior power in putting the elements in order.

The part of Mach's theory which is evident is this: theories, and with them theoretical entities, carry out what may properly be called correlational functions. What is not so evident in Mach's theory of the correlating hypotheses and hypothetical entities is whether they exist in any sense, even if only in our understanding. To be sure, he does not deny the existence of theoretical entities, as any other entities. He only stresses that anything which exists, exists in only one sense, namely as a construct of elements. He puts no special requirement upon hypotheses, but not surprisingly he analyzes the meaning of a hypothesis in terms of human experience.

As a result, we are left with an ambiguous status for theoretical explanation. If a hypothesis is 'relevant' to experience, it will be allowed. But, he adds[13],

> In a complete theory, to all details of the phenomenon details of the hypothesis must correspond, and all rules for these hypothetical things must also be directly transferable to the phenomena.

Here we must protest that detailed matching is too severe a requirement. For Mach's own standards to be met, it would seem sufficient to demand that the hypothesis be set forth in terms derived from mathematics and from the senses, by analogy and extension if not by direct abstraction. It is only when all conceivable experience is irrelevant that a hypothesis should be barred from science: all other hypotheses may be entertained. But still we are left with Mach's phenomenalist view of existence: permissible hypothetical entities and properties, whose existence is postulated by accepted hypotheses, exist "only in our minds" (meaning, as constructs) while the elemental facts which they explain constitute what there really is, the irreducible. Existing in our minds, they are fictions – after all is said and done. Nietzsche's remark is apt[14]: "Parmenides said: 'One cannot think of what is not'; – we are at the other extreme, and say, 'What can be thought of must certainly be a fiction'."

But Mach defends himself. His defense is simple. It is illicit to demand a conception of existence from a theory of science. We can at most ask what is scientific explanation. And for this, Mach offers the observation that scientists explain by a joint procedure of analyzing and reducing[15]:

> Natural science has to resolve the more complicated facts into as few and as simple ones as possible. This we call explaining. These simplest facts, to which we reduce the more complicated ones, are always unintelligible in themselves, that is to say, they are not further resolvable. ...
>
> People usually deceive themselves in thinking that they reduced the unintelligible to the intelligible. Understanding consists in analysis alone; and people usually reduce uncommon unintelligibilities to common ones. ...
>
> What facts one will allow to rank as fundamental facts, at which one rests, depends on custom and history.

Again the culture-bound contingency of the empirical base compels Mach to transcend his own positivist reconstructions! Unlike Hume, who reduced causal analysis to habit and custom and then left the melancholy of his study for the zest of daily life, Mach saw that a theory of history must supplement epistemology, and perhaps even complete it. Curiously for Mach, it seems that ideas of nature must, in the end, follow upon experiences in history. He flirts closely with a Collingwoodian framework of interpretation [16] so to speak; but whereas Mach so thoroughly mastered the history of ideas of nature, at least as they have developed throughout European history, and so little pursued philosophical analysis of historical knowledge, Collingwood suffered and gained from a reciprocal set of neglect and emphasis: he was a great historian of culture but knew too little physics. Mach's 'common unintelligibilities' vary over the centuries, and it is open to the scientific historians of custom and of epistemology to record, and, dare we say, to explain their changes.[17] Were such historical knowledge available, Mach might foresee a choice among alternative basic sets of elements. Within his own actual analyses, however, only a narrow meta-choice is open: whether to accept the experiential base from among the 'common unintelligibilities' of daily life in our culture, or to choose a base by conventional agreement among scientists, or another elite. But naturally Mach will make even this choice among epistemological frameworks on grounds of utility; and he sees it as a sign of scientific progress that such a choice may be considered. And, we should never forget: Mach sufficiently clearly adumbrated the idea of choice to have encouraged the young Einstein to endeavor to create a set of radically new alternatives.

III. SCIENCE IS DESCRIPTION

"Explanation is nothing but condensed descriptions." With this innocuous sounding epigram, Mach states his theory of science. Science describes. Whether one agrees or not, one must admit: how simple and seemingly unproblematic. But how does science describe? By formulating functional relations among recurrent phenomena. Recurrence, we recall, is but a brute fact. Analysis is, too, and mathematical formulation. So laws of nature are short-hand summary conjunctions, in a mathematical language, of their own myriad exemplifications and individual descriptions. Still, all is quite simple and straightforward – suspiciously simple. Why is our suspicion aroused? Because we must ask at once: where do predictions come in? If science is but condensed description put mathematically, there is no room for prediction in science.

And so we will devote this section to Mach's attempt to make science predictive – as indeed he must do – and to a critique of his attempt, and to a Machist rebuttal. The initial step is taken, as may be expected, by further analysis of scientific laws.

Laws are linked with prediction via contingency. They specify the conditions of their own applicability. To Mach, laws are wholly to be understood as conditional prescriptions upon our experiences: typically, he wrote that "if specified conditions are met in experience, then other specified resulting experience will be met". So science is not merely a vast, though condensed, record of experiences. It ranges beyond that record to describe possible experiences. What does 'possible' mean in terms of experience? Mach did not explicate further, nor did he find any evident difficulty with hypothetical statements or contrary-to-fact conditionals. For him, science is meaningful when in use; not when its use is speculated upon. But his theory of science enlarges the extension of 'description' nevertheless. Laws go beyond the given experiences; laws function not only as descriptions but also as prescriptions for descriptions. Mach wrote: "Natural laws are restrictions which, in the light of experience, we prescribe for our expectations of phenomena...."[18] So laws are condensed and extended description. Indeed, Mach was right to say that the condensation effect gives power to science; but nonetheless condensation itself is hardly descriptive of any experience. Yet, condensation and extension carry the gist of his doctrine, and so right away the impression is reinforced that once again Mach simply and acutely sees what the scientist does, but he cannot explain why it succeeds.

Let us begin again. "Explanation is nothing but condensed descriptions." Instead of asking how this takes place, we ask: descriptions of what? Mach's answer: of experience. But, with him, we will agree that experience need not be *past* experience (as in most older empiricisms). So let us observe again what is the relation between experience and theory, this time alert to a more comprehensive restraint upon what Mach would, and would not, consider to be experience. Theoretical statements, as Mach construes them, should wholly be descriptive of experience; they must be analyzable into repeatable distinct experiences, namely repeatable, individual phenomena; they summarize statements of such phenomena; they are tested only by statements of the phenomena; and they are communicable only through public understanding of statements of the phenomena. But while these theoretical statements are so tightly bound to clusters of experiences, they are *not equivalent* to experiential statements. Even Mach, indeed Mach above all others, recognized the outstanding merit of theory, and of entire scientific advance, to be an enormous *elimination* of experience. It is not going too far to say that the true function of theory is to "save the trouble of new experiences".[19] It may sound blasphemous for an empiricist to view experience as trouble, and even as dispensable. However, once again, we may be obstructed by the usual view of Mach as a traditional empiricist. Mach's point, however, is that whereas the old empiricists cherish experience, and package it neatly in laws so that laws would be "nothing but condensed descriptions" of *past* experiences, for him laws are recipes for producing individual descriptive statements. Scientific education leads not to knowledge of past facts but to knowledge of recipes for producing them. That is to say, we have recipes for how – using several of Mach's phrases – to "reproduce in thought all possible" observations of correlations. Our analogy to the kitchen is fair because recipes are to be practiced; when merely contemplated, they demand a full imagination; when practiced, they produce. And so with all formulas. For example,

> Knowing the value of the acceleration of gravity, and Galileo's laws of descent, we possess simple and compendious directions for reproducing in thought all possible motions of falling bodies. A formula of this kind is a complete substitute for a full table of motions of descent, because by means of the formula the data of such a table can be easily constructed at a moment's notice without the least burdening of the memory.[20]

What is known, when we know a scientific law, seems, for Mach, to be a device for calculating, and perhaps in other ways, too, specifying individual states of the experienced world. His program for philosophical analysis of science is thereby close to seeing it as a programmed high-speed computer. Whatever can generate tables of correlated facts will save finite brains the vast and impractical task of accumulating the entries. Indeed, suppose we had the data; then, as we are now no longer surprised to read, Mach said, "If all the individual facts – all the individual phenomena, knowledge of which we desire – were immediately accessible to us, a science would never have arisen."[21] To conclude, science rectifies the weakness of the human body as a system of slow and relatively insensitive observation devices, and of the human mind as a finite information bank and retrieval system.

Condensed description of experience, then. What else? It is not a bit strange to discover that science, as Mach describes it, while seeking descriptions rather than explanations, fails to seek truth. It does, of course, provide true descriptions: after all, it is a mental machine for doing so. But beyond descriptions, it does not seek true explanations of them. Rather, beyond descriptions, science seeks economy of description, not just ease of description. This should be stressed: economy is not mere ease. What exactly is it? Mach's specifications for measuring economy were various, and his terminology is now commonly understood and used: science, he says, seeks simplicity of description because the simpler of two theories is alleged to be more powerful as an information system and as a predictive technique. This does not, however, settle the matter. Alternative theories may be evaluated empirically with respect to predictive power, to be sure, and evaluated logically with respect to descriptive simplicity. But the two evaluations may disagree: Reichenbach once distinguished inductive simplicity and descriptive simplicity[22]; Mach did not do so. Since the concepts do not coincide in intension, they need obviously not coincide in extension.

However this may be too fine a distinction. Mach, we remember, was reporting: his theory of science is descriptive, just as science is, he says. Perhaps economy in action will prove Reichenbach's distinction irrelevant.

Perhaps the point most apt for us to report is the controversy over atomism. First of all, controversies ought to be resolved by Mach's criteria of his tough anti-metaphysics. Second, he was a seasoned party to the dispute, and a wrong party at that. Was his wrongheadedness a proper application of a wrong theory, or was it a misapplication of a possibly right one?

The controversy was formulated traditionally, and the question asked: are there such things as atoms? Now, if 'things' is commonly understood, then Mach could not join the controversy in the first place. If however 'things' are Machist, the answer, naturally, is yes! But his life-long oscillations, between acceptance and rejection judgments of atomism, suggest that for him the question was, or should have been: is atomism a useful doctrine? The answer, we know, is yes. In which case, accept it and use it, of course. But from this we normally and usually go on to ask: do atoms really exist? Not if they take no part in observational experience, says Mach, for whom real existence is reserved for elements alone. But then, we retort to Mach, need atoms – in order really to exist – take part in experience *as atoms*? Mach's reply to this and similar questions is consistent throughout his critiques. The idea of atoms serves a useful pictorial function for those who require such pictures, and he always approved its "heuristic and didactic value". As to whether *science* has to accept atomism, that depends upon whether atomism is not only of heuristic value but also economical; and on this very point, Mach changed his mind. As to the further question whether, assuming that atomism is scientific, we should accept atoms as really existing, Mach's answer is no; the only really existent elements are sensational, experiential.

We should stress the distinction between the heuristic and the economical, the personal and the impersonal. When pictorial thinking is no longer required of modern thinkers, it should be possible "to extract ... the essential factual kernel and get rid of the superfluous accompanying pictures".[23] But then we may as well say 'yes' to the question: do atoms exist?, for, to Mach, the factual kernel was, of course, in genuine experience, throughout the decades of physical and chemical disputes. We can inquire even more sharply: do theoretical entities exist? And Mach's answer will be the same. Seek the 'factual kernel', and tough-mindedly disbelieve both pictorial fantasies and metaphorical fictions. With a theory of science as utility, Mach need say no more. If he were *dogmatically* opposed to the acceptance of the independent existence of atoms, as Boltzmann ruefully characterized him, he was after all only consistent with his critique of such apparently less theoretically formulated concepts as temperature, mass, and space. What *is* independently existing in these – and in the objects of everyday life – could also exist 'independently' as atoms; but no more: namely, publicly and impersonally knowable correlative-functional dependencies among phenomena. From this point of view, Mach had only a technical quarrel with the atomists: was atomism more economical?

And so our example – Mach and atomism – is not sharp enough to make us wonder as yet: how exactly does one measure economy? what exactly is the difference between economy of description and ease of description? is there a significant difference between economy measured as simplicity measured inductively and economy measured as simplicity measured logically? These questions should be answered in order for us to be able to judge whether Mach did, in fact, succeed in incorporating prediction into his theory of science as intended, or not.

Now whether or not Mach is successful, at least we know his line of attack: laws are recipes, instruments, and for prediction among their other purposes. Possibly we may be able to repel this attack on general grounds, and spare ourselves the trouble of detailed argument. It will soon become clear that general considerations are plausible, even convincing, but not compelling. However, I want to introduce such general considerations because they will be useful when we return to the detailed questions just raised.

Did Mach admit hypotheses or reject them? In restricting the cognitive reach of science by saying it gives no explanation of phenomena, Mach was, as we have noted, merely reporting. But he claimed also to be on stronger philosophical ground. He was sceptical about conjectures which are said to apply beyond experience, and this scepticism led him also in principle to reject any hypotheses which apply both within and beyond experience. As may be understood, such rejection provoked a massive critical reaction. Never mind Mach's Bellarminean acceptance of hypothetical formalisms. 'Realists' argued that science gives (partial) explanation of phenomena by framing hypotheses about entities and processes whose properties include interaction with our instruments and with our sense organs but are not necessarily exhausted in these interactions. Mach responded to these critics. He correctly pointed to the primitive role of experience and practical activity in initiating scientific inquiry; and further, still correctly, he stressed the merely convenient and conventionally arbitrary status of many concepts and relational components in scientific theories; and also, once again correctly, he emphasized the absolutely crucial role of utility-in-experience in giving grounds from which to judge (confirm, refute). But Mach gave no convincing argument to suppose that the *evidence* for what we know is everywhere identical with the *meaning* of what we know. "Perceivability is indeed the *criterion* of existence, but not its *definition*", wrote a critic a few years later.[24]

Since Mach recognized no convincing argument on the other side to establish such a distinction, the dispute was not resolved. But Mach's defense transcended any bare utilitarian thesis. He backed his position with a *theory* of utility; and it must be observed that Mach's theory of utility is more subtle, more utilitarian – even more realistic – than his realist critics generally have understood. Mach could be realistic *because* of his account of instruments, not despite that account. Neither meaning *nor evidence* is relevant to instrumentalism. Thus, taking as an example the Mach-instrumentalist view of theories in science, we must grant that our previous account of his view of them was misleading. When, in Machist manner, theories are conceived as instrumental, they are *not* to be thought of as mainly devices for clarifying, arranging, correlating, re-arranging, and otherwise *manipulating* phenomena, nor are they merely logic machines which process observables as input, and yield, as output, accounts of measurements. Theories, as instruments, are not meaningful; they are not abstractions from meaningful basic (elemental) statements: they are plain rules. It follows that theories cannot be refuted.[25]

At this point, Mach's epistemology appears ready to collapse: this theory of science is not, as intended, a theory of cognition but a practical social psychology of the behavior of scientists *qua* scientists. As such, Mach's theory turns onto another route. It may seem to be a retreat, but it is, however, of distinctively philosophical interest. It came about first through a personal chance encounter. Under the influence of his friend, E. Hermann, a competent political economist with mathematical training, Mach early had been impressed by the advantage science presented to the human species.[26] For Mach, science saved thought, as we saw above. The economist almost casually notes that the problem of understanding science is much like that of understanding the growth and use of financial capital. Mach praised Hermann's own formulation: "Science has a problem of economy or thrift."[27] How deal with thrift? How measure it, whether in the market or the laboratory? The measure, it seems plausible to consider, is the efficiency of an expenditure, be it an expenditure of money or an expenditure of energy and thought. What does it do, and how much is the profit per unit invested? Scientific thought is efficient, in Mach's historical view, because it has greater success in adapting man to his environment than shown by most creatures to theirs – although comparative evaluation measures were not needed for his judgment.

Thoughts may adapt to facts – or not – and this is, for Mach, the scientific content, the 'factual kernel', of traditional philosophical discourse about truth as correspondence. Thoughts may adapt to other thoughts, and this is the kernel of the traditional view of truth as coherence. Observation is thought adapting to fact; theory is thought adapting to other thoughts. Either adaptation is biologically efficient. And with this appraisal, Mach has turned to a biological theory of science. He sees thought as a specific kind of biological reaction to another biological process, namely as a reaction to sensation. With classic 19th-century biological modelling, Mach believed that sensations, in the human animal, may (among their other effects) release a reaction whose accompaniment is the adaptation of ideas to facts. When successful, such 'mental life' is an instrument of biological survival.

How amazing to see the metamorphosis of epistemology into biology, just when we have gotten used to the metamorphosis of science (meaning, as usual, physics but necessarily including psychology, biology, and any other empirical study). One might say that Mach transforms physics into the psychology of sensations, and this into biological adaptation: physics to psychology to biology. But these transformations are reductive, not so much philosophical as scientific. However, if his ultimate theory is truly a biological monism, we should reformulate our questions: do predictions help us survive? do hypotheses help us predict? Does metaphysics?

It is easy to answer these questions. Even metaphysics may be favored if it is heuristically valuable, though only to that extent.[28] Our ease in answering from the standpoint of Mach's biologism only makes us suspicious once again; is his biologism not inconsistent with his theory of elements – his sensationalism? Once again, let us proceed slowly.

In order to examine his consistency, we may try to translate his sensationalism into his biologism *and vice versa.* Successful adaptation to facts is realized in practice by anticipation and prediction of facts-to-come, experiences-to-be. Moreover, successive enlargement of the factual environment, in the course of history, imposes a further requirement upon ideas: they must be transformed in order further to adapt. Mach soberly concluded that "... all depends upon the capacity of adapting existing notions to fresh facts".[29] The dialectic of old and new ensues for Mach's view of science. As we remember, the certainty which he ascribes is not to the results of scientific investigation but to its (dialectical) method. In the dialectic, knowledge plays both roles. Knowledge of correlated phe nomena characteristically leads to expectations, habits, and predictions in daily life as in science; and it leads to pre-judgments (which are judgments which we make without reiterated testing) and to prejudices. What shall we make of knowledge as prejudice? Mach's view is an insightful play with words. He emphasized that individuals and societies cannot live by constant tests: literally prejudicial practices are literally essential to life. It follows from this that prejudice is to thought what instinct is to physiology – and Mach cleverly called prejudice "a sort of reflex motion in the province of intelligence".[30] Then – and here Mach is superior to Santayana's scepticism – the powers and the illusions of prejudice are assailable by further critical adaptation of thought to new facts, not by rejection of prejudice.[31] Not science but practice, crashing down upon ossified thoughts, can force a conflict with prejudice to a resolution.

In history, the struggle of ideas for adaptive survival have been as severe as the struggles for survival of whole species. Ideas compete; some will be found to be flexible, others rigid. The horse changed, and survived; the dinosaur held fast, and died.[32] Moreover, the struggle is not only to adapt, but to adapt more effectively than do competitors. Without doubt, Mach felt, science is a marvelously deliberate method of choosing ways to adapt to the environment; and to adapt our chief mode of adaptation – which is thought – to further changes and new experiences. Hence, science is the destroyer of habit and prejudice, but likewise their victim. It is hostile to any belief in miracles, indeed science dis-illusions and dis-enchants, for illusion, in Mach's judgment, is fundamentally counter-adaptive. And yet, even in their downfall, Mach sees a positive role for illusion and prejudice: "In the struggle ... between judgment and prejudgment (prejudice) ... our understanding of things broadens."[33]

This much is science, seen biologically. But what is biology, as seen from Mach's sensationalist theory? Evolutionism is a theory; adaptation for survival is a part of that theory. Theories are condensed descriptions of experiences. Experiences are transformed by this analysis from ego-centric to impersonal. Perhaps experiences are not necessarily *past* experiences; but how may they include all the apparently pseudo-experiences? Future experiences? May theories refer to hypotheses of trial and error? Ideas can, of course, be false, but how condensed descriptions? How, in general, may we explain adaptation of ideas? In a broader context, if ideas compete, we have richness; whereas Mach's original program was austere. He knew it. "... natural laws are restrictions which, in the light of experience, we prescribe for our expectations of phenomena ... in place of description, I propose the expression *restriction upon expectation*, so as to show the biological significance of laws of nature."[34]

Thus there are two Mach's: one who sees an important role for hypotheses – even when they are prejudices – and one whose program was to eliminate from science all illusion and pretense, to wit metaphysics. The reason that the anti-metaphysical Mach did not feel constrained by the fact that science is drenched in hypotheses is that, when facing this, he played the evolutionist Mach.

IV. UNITY OF THE SCIENCES

We have come to a total collapse: Mach's sensationalism is austere whereas his evolutionism is prolific. Were this the last word, Mach would yet be an intriguing and important figure. However, it is not the last word. Surprising as it may sound, I wish to suggest, what Mach would surely accept, the saving role of his thesis of the unity of science. To unify the two Mach's is to unify two disciplines. It is not to reconcile two philosophies. Hence, the divergence in his two scientific theories about science threatens one of the most prominent themes in positivism, whether Comtean or Machian, and certainly the most publicized aspect of the 20th-century Machists: the unity of science. If we can defend the conception and program of this unity, we may be able to return to defend Mach's entire outlook. But it will transpire that his 'unity of science' is not univalent in its meaning, and – less surprising – that its ambiguity both carries on the previous tension in Mach's analysis, and also suggests a decisive and transcendent cure of that tension through an ancient positivist metaphysics.

Unification of distinct realms of knowledge is not a new goal. In many epochs, thinkers try to see the world as a whole, to struggle with observed co-existence of incompatible realities which yet are somehow compatible with the same space or time or body or mind. Such unifiers find the diversity epistemologically contingent; they cannot believe that it is ontologically so, and this irritant disparity between knowledge and belief leads to the program of unifying the sciences. The belief is frankly speculative[35], and yet it surely is programmatic for science; nevertheless, while often metaphysical, it is not necessarily metaphysical. Indeed, since a major part of scientific motivation, whether instrumentalist or cognitivist, is served by finding connections and regularities where there had been disconnections and chaos, the unity of the sciences can be formulated as a scientific hypothesis[36]; and it can inform crucial tactics and strategies of the various specialties. Several theories of unity of the sciences have been offered, each of which is relevant to Mach's philosophy. First, there may be a basic science to which all others may at some time be reduced and which may describe (or explain) all phenomena, at least in principle – reductionism; as, for example to Mach's elements. Second, there may be fundamental methodological unity among the sciences, with diversity of subject-matter; in particular, procedures of display, communication, confirmation, and measurement may be identical, although conceptual reduction is not possible – unity of method; as for example Mach's sensationalism. Third, there may be a generally valid theory of the different sciences within which their differences become plausible, and this theory – as itself a scientific description – in a sense unifies, in the sense in which, for instance, Mach's evolutionism regards different disciplines as different adaptations to different environmental factors. Now, this theory – the science of science – has, naturally, its own status as history, or psychology, or logic and it may raise subsequent puzzles.

How come Mach participated in all these trends? My answer reflects his appreciation of the several sciences, for each is linked with differing attitudes toward the problem of unity. To Mach, as to many other scientists, it was evident that there are alternative directions of investigation to the same matter. This is reasonable, and even common: there is the physics of falling bodies, and it applies to falling organisms as well, but there is the chemistry and biology of those organisms, and indeed possibly the psychology and sociology as well as other disciplines. If a falling body seems to disobey Galileo's laws of free fall, then perhaps the true description will modify 'free' by conditions which are perceived through the other sciences. Not that psychological beings can violate the physical laws, but only that they may utilize those laws for their purposes, as in flying.

Here Mach's theory of having no distinction between ego-phenomena and physical-phenomena pays handsomely. We may indeed reformulate it to say there is no distinction either from a psychological or from a physiological viewpoint, and the old point covers the new: the phenomena – as 'sensed', and this term itself may be one in psychological or physiological language – are grouped for convenience, a methodological criterion with no direct implication for the possibly diverse characters of the traditionally subjective ego-phenomena and traditionally objective physical-phenomena. Rather than such an implication, Mach drew his neutral interpretive picture, as we know, and called these experienced phenomena 'neutral elements'. He has, therefore, an explicit unity of all types of phenomena because of this fundamental doctrine of experience. He does not claim any decisive philosophical victory, however; in fact, it might be said that the unity achieved is so bare, so minimal, that the principal questions of scientific as well as philosophical understanding remain unanswered. I have in mind, particularly, whether the world is so made, that reductionism – any reductionism – is true. But Mach's achievement is a little more substantial. The common experiential basis, interpreted – reductionistically perhaps – to consist of elements, is more than a primitive starting-point of a theory of scientific knowledge; it is also – we remember from his evolutionism – the foundation for an empiricist doctrine of confirmation or refutation of all cognitive statements, and of the connected empiricist doctrine of rejection of non-cognitive statements. Hence Mach's self-appointed task of purging science is *not*, as many have understood, a priestly and self-righteous task of repelling infiltrating heresy; it is *not* scientism; rather, it is a perennial task characteristic of scientific evolution as such. Mach need not, and did not, insist upon a reduction of the various scientific disciplines to a simple and common content of homogeneous sensations, but he did insist upon their common test criteria in human experience. The totality of human experience was admitted; and he offered the striking hypothesis that all experience might be recognized, in principle, to consist of elementary simple bits. Given his hypothesis – said to be a description which would be evident upon examination of various experiences – one may turn to the complicated task – the perennial task, indeed – of showing how knowledge of apparently complex entities and processes may be translated into knowledge of elementary received data.

Again, it is from the vantage point of his hypothesis that Mach has proposed a language usable for all the sciences – as a program for the reduction to elementary data and to the extent that it can be achieved. Such a program is put somewhat more formally into syntactical procedures whereby sentences about entities and processes (whether denotative or theoretical) can be replaced by sentences which refer solely to elements.[37] What is central here is that the elements are contingently given, or received; they are received as, and constitute, experience; and they come without necessity or any other rationale. To refer to them as 'sensations' is tentative and risky. Why? Since the subjective, personal and apparently private associations of that term are evident. Since Mach's understanding led him from elements of reality and knowledge to a public and impersonal conception of the world. The elements provoke a revealing query: is his conception of elements equivalent to any doctrine of atomic experiences, or atomic sensations, or atomic entities, or atomic facts and atomic sentences – namely, the world of logical atomism? Is Mach as Machist as, say, Russell of *Our Knowledge of the External World*? If so, what a curious situation for the leading opponent of chemical atomism. In fact, however, Mach did not offer more than a tentative theory of the nature of the elements, without Hume's or Russell's explicit attribution of ultimately atomic and irreducible status to them. He did not claim final knowledge of whether experience be continuous, or atomic, or dialectically changing; he has no settled knowledge of any kind; he would have considered the finality of sensa to be a question for further empirical investigation. But whereas finality is still in question, the empirical unification of science has already been settled.

(So, my reading of Mach certainly differs from that of Russell. Which is correct? What characteristic of the Machian elements may, possibly, arbitrate between the two readings? Aside from being neither mental nor physical in the usual usage, the elements are subjected to logical and mathematical arrangements which also do not characterize the elements themselves. Whichever way the scientific and philosophical status of logic and mathematics may be otherwise conceived, at least in Mach's conception they are no more arbitrary, no less real, than are the laws and entities of physics and biology, nor *a fortiori* those of psychology. Mach's positivism, therefore, is consistent with logical realism as much as (or as little as) with naturalism.[38] And so, clearly, Mach's logical realism is inconsistent with the epistemic and reductionist sensationalism of Russell.)

We have, then, a phenomenology of an uncommitted sort, still open to several interpretations of various of its aspects. Mach offers injunctions for scientific study of the elements, derived from his historical and empirical research efforts, generalized, that is to say, from what scientists have done. To the rule of economy of thought, he adds his most important conception of investigation; "... the principle of the complete parallelism of the psychical and the physical ... which recognizes no gulf between the two provinces...."[39] This does not mean mere correspondence of a physical entity to every psychical (and vice versa, Mach adds! and adds that this mutual correspondence is generally believed), nor a double-aspect theory, whether as in Hume or Spinoza or even Fechner. Admittedly, the elements are neither psychic nor physical; only our conventions of grouping them may be so labelled. Nevertheless, from Mach's most interesting and most detailed examples, we may conclude that he causally relates psychological sensations to physiological (i.e. physical) accompanying conditions and processes; and, even more, at a later passage, he notes, like a good reductionist, perhaps even materialist, that his "ideal of psychology is that it should be purely physiological"; and yet, at the same time, he urges that it would be "... a great mistake to reject so-called 'introspective' psychology entirely ... in many cases the only means of obtaining information as to fundamental facts".[40] Mach offers a hint about his preferences: physical description is more precise, more completely determining, whereas psychological description is vaguer, more fragmentary; indeed, for Mach, the psychical experiences are determined by physiological circumstances which are – at least for present scientific knowledge – too complex to be understood, and, quite independent of whether we ever achieve complete descriptive sciences, "... only a very minute part of the traces of physical processes ever comes to life in our presentation". It is in the 'presentation' that Mach, and we with him, live our experienced lives; but as scientists we may bring forth, from its unexperienced underground role, the physical and physiological fact "... from its isolation...". We bring the physical and physiological into our psychic life, as a crucial element of experienced reality, and we perceive it scientifically as indeed physical and physiological, i.e. we perceive it within the flux of experience as scientifically and descriptively correlated, and doubly so: correlated systematically with other physical and physiological facts (elements), and correlated to an increasingly clear extent with psychological or mental phenomena. To make this point, let one draw an analogy: the determinism provided by the unconscious in Freud's attempt at an explanation of conscious experienced life was similar to the determinism provided by the physical and physiological in Mach.[41] The analogy, incidentally, is no accident: Freud had read his Mach carefully. "The science of psychology is auxiliary to physics. The two mutually support each other, and it is only when they are united that a complete science is formed", says Mach, and Freud would of course concur.

The prospects of such unity vary from time to time. Unity of science was not a fact then, when Mach wrote, nor is it now. His hospitality to theory seems to provide him with a Machist optimism. We recall that, for Mach, theory need not be pictorial. It need not, going a bit further in Mach's spirit, be analogous to the macro-observables. The phenomenalist Mach, in a wry and profound way, cherishes sound theory; he places no limit upon the patterns of conceptualization, nor any Cartesian bifurcation between empirical realms. The unity of science movement[42] failed to enrich science as Mach ultimately hoped. But his heirs did what he prescribed to be done: common methods of test, shared goals, joint possibilities of utility, the social function of disillusions, mutual supports, encyclopedic integration. Behind these, a genuine Machian unity of description was hoped for, but the enrichment was lost on the way. In a most touching passage Mach said that the scientist's thoughts are not outside of nature; as the scientist is part of nature, his thoughts are too. And the natural world is not split apart; it is one. This is much more than mere physicalism.

The question still remains, an ontological one: how can Mach see things and thoughts integrated?

V. THE HUMAN SPIRIT

"The highest philosophy of the scientific investigator is to *bear* an incomplete conception of the world and to prefer it to any apparently complete but inadequate conception."[43] Mach was a positivist and an historian. Each contributed to his description of the motivation and dynamic of science. They are linked by his instrumentalism. But in two respects, Mach's instrumentalism is incomplete – or incompletely formulated.

First, instrumentalism is criticized as irrational, most notably in Mach's context by Husserl as early as 1900.[44] How, asks Husserl, can the descriptive principle of economy of thought be taken as a rational foundation of the structure of scientific knowledge? The principle cannot provide a basis for rational explanation, such as Husserl, and others, see in mathematical laws or axiomatized laws of physics. Is not Mach's economy of thought just an arbitrary teleological statement, not necessary, not explanatory, not rational? Mach replies, with disarming, indeed alarming, modesty[45], that "a theory of theory was too difficult a task" for him, that he had merely directed his attention to several individual phenomena of the scientific enterprise – adaptation of ideas to each other, intellectual experiments, continuity of thought, and so forth; and he reports it was helpful to construe science as an evolutionary phenomenon within the course of biological nature "... in which logical thinking assumed the position of an ideal limiting case". No doubt, he admits and welcomes, purely logical analysis of the sciences can be carried out, and should be. However, "even if the logical analysis of all the sciences were complete, the biologico-psychological investigations of their development would continue to remain a necessity for me", and then he adds with good humor that nothing "... would ... exclude our making a new logical analysis of this last investigation". And he mischievously implies that there is a need for historical-psychological investigations of logical analysis itself. Nevertheless, the point against him is made: simplicity and economy of thought were teleological or regulative principles, not 'justified' by reason, imposed upon experience only by practical considerations; and Mach can only agree, and urge that any impartial observer will see that this is how science does its job.

Then, what really is that job? to investigate the facts, the elements, and thereby help men to survive their environmental troubles; or, far more ambitiously, to change, modify, re-arrange, and even newly construct the world? Much later, Husserl[46], among many others, pessimistically saw modern science as a super-instrument, made for a vast successful project: to transform our world by mechanism to a mechanism. Mach could not rebut; he was a social optimist, who would have been profoundly unprepared for the efficient barbarism of the new century; Mach characteristically found science part of a progressive enlightenment and gradual amelioration of human life in the given world of experience. When science goes beyond adaptation to construct a new human environment, to make a new world of experience, instrumentalism must be subordinate to a deeper, or a larger theory; Mach was silent.

Instrumentalism is rooted in a theory of choice which is dogmatic: choose what is useful. It is dogmatic since, in order to be undogmatic, the injunction – or recommendation, or description of accepted recommendations – would have itself to be rooted in a context of discussion of alternatives to it. Mach did not sketch such a context, but he was forced, by this apparent arbitrariness of the instrumentalist theory, to enlarge, and ultimately to change, his theory of science. Science enlarges the private realm of each investigator far beyond his personal sensations and ideas; it enlarges the spirit by developing an *enjoyment* of understanding. Scientific thought "... creates its own ends, seeks to satisfy itself by suppressing all intellectual uneasiness".[47] Born of need in prehistoric practice, as Mach well knew and loved to investigate, science developed to become a historic factor which liberates as well as protects. Liberation is possible, and protection too, because of the peculiar creative role of thought. We bring a degree of arbitrary unity to the world of disparate elements, or so it seems to a naïve positivism. And yet, to Mach, the elements were not mutually incompatible, not chaotically formed, not undependable. Without a theory of mind, nevertheless, he would agree with the aphorism of the psychoanalyst Ferenczi[48]: "Sense organs separate the world into elements, the mind restores it."

May we contemplate the growth of science biologically as the development of a new organ, the mind, and at the same time understand the behavior of the mature mind psychologically through its origins? First, of course, we must ask: which origin? Tribal or individual? No matter which, says Mach, since their developments are the same. Mach was convinced that the history of science in primitive times corresponded to the spontaneous technical responses of very young children; he used his son Ludwig as a test of this. His technique of investigation was not direct observation. It was an attempt at recovery of infantile memories; these revealed what Mach described, in his last work[49], as "... powerful, inextinguishable dynamic sensational experiences (which) transport us suddenly nearer the instinctive origins of all contrivances such as tools, weapons, and machines". And later, in a passage about primitive technology, he observed: "If we *think and dream*, however, about these things of long vanished ages, then, like illusions, old memories of experiences and feelings waken in us, and, sinking back into our once childish world of sensations, we dimly apprehend and await the manifold developments and means of discovery of those contrivances of such immeasurable range." These are statements Freud could make: the continuing reality and efficacy of sensations which are no longer sensed, neither directly nor in memory, but repressed[50]; the role of dreaming as an entry to a part of the world of experience which is not publicly observable nor directly introspected; the search for culture history through individual psychology. And Mach's most poignant remark in this connection: "It may well be that unconsciously sensation and understanding have their roots in our or our ancestors' memories." Ferenczi, that most philosophical of psychoanalytic therapists, welcomed these speculations of Mach, and urged collaborative research upon the genesis of scientific activities by psychologists, physicists, and other specialized scientists.[51] Although respectful, Ferenczi criticized Mach for failing to comprehend the libidinal root of scientific and technical discoveries; although he agreed with Mach concerning adaptation to reality which is the other root, he regretted that Mach did not fully perceive the libido within science. In other words, according to Ferenczi, there are two sources of human development (science included), adaptation and libido, reality and pleasure; and Mach updated would say only adaptation, with libido as its instrument. Which is right is an empirical question. Adaptation to reality is, to the psychoanalyst, 'egoistic', and should be linked with the 'ego', to wit, with rationality and intelligence.[52] To Mach, in contrast, adaptation to reality is non-egoistic, indeed, leading us to no ego whatsoever, in spite of its having become rational and intelligent in mankind. Ferenczi did not note how substantial is the challenge of Mach's psychoanalytic hypotheses to Mach's logic of science. Here, in another form of biologism (which is to be sure also evolutionary, as before), Mach seems to find the causal emergence of an objective ego, whose characteristic experiences have a coherence and a deterministic structure – indeed a mechanism! If the theory of neutral elements were ever to transcend the egocentric precipitate of sensationalism, and ever to avoid the metaphysics of material and independent things-in-themselves, it now would have to do so by working through the analysis of the human spirit as both determinedly egoistic and objectively independent. This one might fear, is just too much to ask. Not because Mach was at the end of his life – every one of these ideas was stated in full or in clear anticipation during many decades – but because he had carried a descriptive instrumentalist view of science as far as it might go.

But he had a way out. It was a classic: a classic of positivism, psychoanalysis, mysticism. Passages from his inaugural address in 1883 as Rector of the University of Prague – indeed from the conclusion – and delivered in mid-life – will suggest his outlook.[53]

Ideas are not all of life. They are only momentary ... designed to illuminate the paths of the will.

The transformation of ideas ... appears as a part of the general evolution of life, as a part of its adaptation to a constantly widening sphere of action.

The arm of man reaches [far beyond the immediate]. What an immense portion of the life of other men is reflected in ourselves; their joys, their affections, their happiness and misery!

How great and comprehensive does self become in this conception; and how insignificant the person! Egoistical systems both of optimism and pessimism perish with their narrow standard of the import of intellectual life. We feel that the real pearls of life lie in the ever-changing contents of consciousness, and that the person is merely an indifferent symbolical thread on which they are strung.

Mach's elements, we remember, are neutral. The same elements, in one or another formation, make up the great world and my self. Moreover, and the point must be stressed, the same elements may make up different selves. To Mach, it is imperative to overcome any idea of fundamental privacy of the individual ego: "... I no more draw an essential distinction between my sensations and the sensations of another person." And immediately he added[54]: "The same elements cohere at a number of points of combination, which are selves." Schrödinger wrote in the same mood, and walking in Mach's footsteps, that "shared thoughts ... really are thoughts in common."[55] And Schrödinger elaborated upon the expansion of the individual as being (or, is it becoming?) an aspect of a greater unity. Through Schopenhauer, through the Upanishads, through Buddhism, the theme of such transcendent unity has served to dissolve both intellectual and personal problems of individual consciousness. Just a few months before that blazing period of the initiation of wave mechanics, Schrödinger wrote two sentences which for all I know might have been distilled from Mach.[56] "The external world and consciousness are one and the same thing, in so far as both are constituted by the same primitive elements. But we are then hardly even using a different formula whether we express the essential *community* of these elements in all individuals by saying that there is only *one* external world or that there is only *one* consciousness."

And so, with this serene multiple ambivalence (are we all one now or not yet? internal or external?), we see the solution to which Mach is drawn. His descriptive science of science has a philosophy to it after all. Science brings liberation in the world of action, we have seen repeatedly. Science also brings salvation to the individual. Can the instrumentalist theory of liberation and the transcendent conception of salvation go together? What is Mach's idea of salvation? Can that salvation be described? Can it be conceived? It rests upon the impersonal nature of the elements which, after all, are of an old metaphysical lineage for they are un-individuated, universals in perceptions. Mach's vision is severe. Science brings world unity, by denying the existence of any true individuals. His vision connects thought to being, by denying any distinction between them. His vision enlarges the experiences of the human spirit by obliterating any more than symbolic existence of that spirit. His vision dignifies man, by reductively dissolving him into a world without dignity or indignity. His vision demonstrates the historical evolution of man (of man's knowledge, power, action, and pleasurable satisfaction), by asserting the eternal world, exhibited even in the unchanging flux of the elements. If Mach's descriptive interpretation is correct, science finally countermands its own very success.

Mach's solution is, however, not required of us. Nor of him. It is only a vision – and I am using this word pejoratively now. It is well couched in an emotionally moving language of self-belittlement, which at the same paragraph however glorifies the self-asserting conquests of nature and the sharing of those conquests. And so the vision is, alas, timid, regressive, metaphysical. Timid as it cannot confront the possibility of the individuality of life, of emerging, developing, unrepressed individuals. Regressive, as it idealizes all reality backwards into the *Urzeit* of psycho-genesis and bio-genesis, to the pre-individuated fluid of ocean and womb. Metaphysical, as it articulates a profound feeling of tranquil self-absorption in the vastly larger world, the deadly 'oceanic feeling' of mystical religion. And it carried out this articulation solely as feeling, without cognitive criteria or practical test – as all mysticism finally must.[57] It will not do. Life is individual, and science is a property of human life. Mach knew this of science, but his vision of the human spirit, though it leaves no room for death, does so at the enormous cost of denying life itself. And thus Mach returned humanity to pre-existence or to lifelessness.

APPENDIX: MACHISTS AND MARXISTS: BOGDANOV AND LENIN

In 1908, Lenin insisted that Bolsheviks must choose: Mach or Marx. Mach's interpretation of science had appealed to many revolutionary socialists in Central Europe – especially in Austria – and it also influenced an important group of Russian Marxists. As Marxists, and as Karl Marx had intended, they wanted to be scientific in their world-outlook, in their social theory, in their conception of history, and in their revolutionary activity. To be scientific included hostility to mystical religion, which rejected rational progress, and to metaphysics, which adhered to absolutes and refused the hard sceptical procedures of observation and experiment: science would end illusions. If Mach could provide a philosophy of knowledge as tough-minded as empirical science, then his philosophical methods must be used as a new basis – indeed as a philosophical revision – for Marxist theory. So far, so good. But the slogan *anti-metaphysics*! has been carried on many banners. It was the particular direction in which the pure empiricist, Mach, marched that Lenin so strongly feared was roundabout; for Lenin thought it led either to complete scepticism, with the result that knowledge of experience can provide no rational assistance to practical endeavors, or to supernatural theology, the abandonment of science: Hume had drawn the former lesson from his empiricism, Berkeley the latter lesson from his immaterialism.[58]

Lenin's rejection of the phenomenalism of Mach was, of course, decisive – politically and also theoretically. It dominates Soviet philosophy. Nevertheless, aside from the strictly positivist core of Mach's philosophy of science, there were other factors in Mach which attracted Russian radical critics, and which deserve some notice. Moreover, the previous century of philosophical developments in Russia had so set matters that the issues Mach was facing, and the methodology he was developing, were appropriate there too.[59] During the mid-century, there had been outstanding thinkers who took the tradition from Hume to Mill, Comte, and Darwin as their own: for example, N.K. Mihailovski, who transformed epistemology into a problem of psychology. And there were others who followed Hegel, then Feuerbach, only to return back to Kant, as so many Central-European Marxists were to attempt; among these was P. Lavrov, who also adapted Comtean positivism to such a journey. Mach apparently succeeded in combining a Kantian appreciation of the active, even constitutive, role of the mind in generating science with a scientific, which is to say, empirical-biological, theory of the origins and functions of the mental life, and specifically of the role of knowledge and ideas. All this – and coming from a highly competent physicist – was dazzling to those who were unable to find in Marx a satisfying Hegelian or other dialectical rendition of science. Many Russians came to acclaim Marx for his scientific account of history, but his views of the natural sciences and of the classic issues of epistemology were nevertheless left open to scepticism, supplementation, and revision.

There was, however, another side to 19th-century developments: a typically Russian apprehension of the dangers of subjectivism; and this fear is the counter-current to empiricism and Kantianism. Both atheists and theologians saw every subjectivist conception of knowledge and of life as a decisive, and irrevocable, step toward solipsism. They perceived the logical outcome of solipsism to be utter corruption: Bakunin said it must come to suicide, Belinsky and Herzen said it brings murder and other crimes as its implication; Soloviev and Dostoyevsky saw both murder and suicide in it. And all felt that what was logical in theory would become necessary in fact. Hence, for many Russians, theoretical philosophy must lead practical life in a great struggle to overcome any private inward focus of consciousness, a struggle toward other persons and toward community.

Was it possible that Mach's theory of neutral elements avoided the danger of subjectivism while guaranteeing the achievement of science? There were those among the Marxists in Russia who thought so, including Lunacharsky, Gorky, and the principal theorist among them, Bogdanov.[60] Bogdanov, who wrote the introduction to the Russian edition of Mach's most influential work, *The Analysis of Sensations*, was a medical and biological scientist. Later, during the early Soviet period, he helped to found the pioneer Institute for Blood Transfusion Research; he died at the Institute in 1928 as the result of an experiment on himself. A hard-working though not eminent scientist, he was also a vigorous philosophical thinker and a minor literary figure. In Bogdanov's philosophy, and in the philosophical writings of other self-styled empiricist and positivist Marxists, Mach's own positivism was enormously attractive because of three principal themes.

First, Mach offers a monistic interpretation of scientific knowledge, which would support a unification of all endeavors in thought and action.

Second, in contrast to unity of thought, every dualism can be seen to be a hidden form of fetishism, an anti-human idolatry.

Third, revolutionary replacement of fetishism in the practical affairs of social life can be prepared only by positivistic annihilation of all fetishism in intellectual matters – whether the fetishism be characteristically animistic or whether it be mechanistic.

It is also true that Machist parallels to aspects of Marxist theory may be emphasized along with the additions to that theory which Mach may bring.[61] Marx, for example, had construed the individual as an ensemble of social relations; and we recall that Mach construes things as ensembles of relations. Marx stressed the pre-eminent role of practice, and of the necessary and laborious struggle for the means of existence, in the generation of human culture – the symbiosis of a utilitarian base and a cognitive or esthetic cultural superstructure – and Mach too found knowledge and culture to have arisen in the struggle for economic existence. Both men also recognized that subsequently science and art have major degrees of independence from economic demands, and indeed that science and art give us the principal ways to transcend brute labor, to overcome private and provincial narrowness, and even to participate in a reality which is beyond the life of an individual or of an epoch. Each learned much from Darwin, although in their readings of the evolutionary doctrine Mach stressed adaptation while Marx stressed progressive and fundamental change. Each of them opposed fixed and absolute notions; and in their separate studies, they combined logical analysis with historical investigation, whatever the subject-matter. And Mach, in personal conviction, was a socialist and an atheist.

Anti-metaphysics, although hardly a political slogan to move the proletariat and the peasantry, was a revolutionary standard within philosophical culture, as Machists understood it. This was disputed; and the argument has not ceased. For if 'anti-metaphysics', as scientism, is linked with scientific domination of men in technological society, then a new and progressive social critique is easily linked to metaphysics, now re-interpreted and conceived as critical social theory.[62] Nevertheless, it cannot be denied that the modern positivist attacks upon metaphysics were loosely associated with revolutionary forces of enlightenment, of rationalism, and of political and economic reconstruction.[63] The supremacy of dogmas and absolutes over the mind, of untested and encrusted superstition over the spirit, of priest over prophet, of institutions over men, of linguistic habit over critical thinking – all these were to be overcome by positivist reasoning. And when such a theoretical achievement was available to a political revolutionary movement, the political gain – or so the Machists claimed – would be more reasonable, more humane. Mach, seen in this context, had worked out a theory to expose the idols of spurious thought. Just as Marxist atheism is a battle against idols and fetishes, so Machist anti-metaphysics is a battle against idol-producing ways of thought.

Bogdanov named his own philosophy 'empirio-monism'. Revolution in the social order would harmonize social experience, and in like manner, empirio-monism would provide a philosophical foundation for scientific revolution in the realm of thought. With Mach, Bogdanov sees all reality as experience, systematized and ordered. Natural science is systematization of technical experience; and experience is made of Machist elements. The physical world, the material reality, is interpreted as "socially agreed upon, socially harmonized, or, in a word, socially organized experience". The psychological world, the personal reality, he interprets as an individual organization of experience. Then, what is true is what is socially accepted in a given epoch, "a mechanism, by aid of which reality is cut up, trimmed and tacked together". He tried to reject all subjectivist readings of his views by pointing to 'universality' as the feature of objectivity; 'subjectivity' is, in this manner, conceived as a character of those Machist elements which are meaningfully organized *only* for one or several individuals. (No explanation of the cause or genesis; just a Machist descriptive distinction.) And so the prime importance of an objective world over any subjective world allegedly is maintained; and it is maintained in a way which Bogdanov felt retained Marxist as well as Machist insights, since the objective physical and social world is produced by the individuals, indeed it is the product of their social organization and of their 'harmonization' of experience through social actions *and conflicts*! What then is the nature of philosophy? Originally a codification of fetishism, philosophy passes, according to Bogdanov (who adapts Comte), from utopian and pre-scientific imagination to a new stage, in which it becomes the general theory of scientific method. From fetishistic dualisms to scientific monism, philosophy is a servant of social powers and dominations; in the new society it will be the guide to efficient choice of techniques and ideas. Bogdanov wrote science fiction in Machist spirit; in a novel of life among Martians, he wrote of a scientist, Xarma – a disciple of Marx –, and Bogdanov has his Martian invent a universal science of organization, a master methodology of problem-formulation and problem-solving via mathematical calculation. Thus, as a technologist, Bogdanov anticipated much, and yet as a philosopher of political change he lost out completely.

Why did he lose? Why was the opposition so vehement? Much depended upon the interplay of ideological practice with theoretical outlook. In the pre-communist milieu, subjectivism had already been linked, indeed almost equated, with scepticism; and would scepticism not destroy a revolutionary spirit? Can subjectivism truly be freed from egoism, and from an ultimately anti-revolutionary position? How can a Machist understand change in human life, for example, change from ignorance to knowledge? and if it cannot be explained but only described, how can Mach describe, and predict, the process of deliberately acquiring knowledge? These, and questions of the ultimate reference of neutral elements, provoked the critics, who demanded a fuller program of philosophical doctrine. The Marxist judgment of Mach, and of Mach's socialist adherents, was bound to be severe. Despite Bogdanov's hopes, the Russian Marxists (and most European Marxists agreed) saw Mach as part of a new tradition from Nietzsche to Bergson and Husserl; it is a tradition which relies upon immediacy and intuition as a foundation of knowledge.[64] Inadequate and incomplete as a theory of knowledge, from the Marxist perspective, Machism was not a guide to life, *and not even a successful 'analysis of sensations'*. The criticism of Mach reached, therefore, into the center of his conception. "Between the object and consciousness sensation stands, therefore, as an impassable barrier. The result is that the influence of the object on our senses, instead of putting us in touch with it, cuts us off from it."[65] Lenin and his followers demanded explanations from science, and from the philosophy of science; they sought perception of reality, not mere sensations, however correlated.[66] Their philosophical intuition was correct, whatever the fate of specific arguments. Mach combined sensationalism, instrumentalism, biologism into a theory of human knowledge. Sensationalism analyzed knowledge by describing. Instrumentalism and biologism explained knowledge by functional analysis but eliminated any theory of objective truth.[66] It seemed that Mach could account neither for active individuals nor a reality within which they act and create. In the end, as we have seen, Mach's profoundest reflections led to serenity and quiet; Machist philosophy could not inspire men in this world of conflict and action.

REFERENCES

*Reprinted from *Synthese* **18** (1968), 132–170.

[1] I am most grateful to my friend and colleague Joseph Agassi for his careful and generous assistance with the final draft of this essay.

[2] R.S. Cohen, 'Is the Philosophy of Science Germane to the History of Science? The Work of Meyerson and Needham', *Ithaca: Proc. Xth Internat. Cong. Hist. Sci.*, Hermann, Paris, 1964, pp. 213–223.

[3] See Joseph Agassi, 'The Nature of Scientific Problems and Their Roots in Metaphysics', in *The Critical Approach to Science and Philosophy* (ed. by Mario Bunge), Free Press and Collier-Macmillan, New York and London, 1964, pp. 189–211.

[4] For a more extended discussion, see R.S. Cohen, 'Dialectical Materialism and Carnap's Logical Empiricism', in *The Philosophy of Rudolf Carnap* (ed. by Paul A. Schilpp) Open Court, LaSalle, Ill., 1963, pp. 99–158, particularly Section 2, 'Kant, Mach and Positivism', and Section 3, 'Solipsism, Conventionalism, Philosophical Atomism, Phenomenalism'.

[5] *The Science of Mechanics*, 5th English edition (trans. T.J. McCormack), Open Court, LaSalle, Ill., 1942, p. 580. Hereafter cited as *Mechanics*.

[6] I am referring here to Fichte's *Human Destiny* without wishing to enter upon a discussion of Fichte's several phases. Even in Heine's penetrating interpretation, in his *Religion and Philosophy in Germany*, where Fichte's philosophy has no self but only thought-processes, there is enough dualism for Mach to abhor – a dualism of elements and of thoughts.

[7] *Analysis of Sensations* (trans. C.M. Williams and revised and supplemented by Sydney Waterlow from the 5th German edition), Open Court, Chicago and London, n.d. [c. 1911], p. 29. (Hereafter cited as *Analysis*.)

[8] *Popular Scientific Lectures*, third English edition (trans. T.J. McCormack), Open Court, Chicago, 1898, p. 209. (Hereafter cited as *Popular*.)

[9] *History and Root of the Principle of the Conservation of Energy* (trans. P.E.B. Jourdain), Open Court, Chicago, 1911, p. 49. (Hereafter cited as *Conservation*.)

[10] *Ibid.*, p. 49.

[11] See the clear exposition and incisive critique of instrumentalism in K.R. Popper, 'Three Views Concerning Human Knowledge', in *Contemporary British Philosophy* (ed. by H.D. Lewis), Allen and Unwin, London, 1956, pp. 355–388; also reprinted in K.R. Popper, *Conjectures and Refutations*, Routledge and Kegan Paul, London, and Basic Books, New York, 1963, pp. 97–119.

[12] *Conservation*, p. 57.

[13] *Ibid.*, p. 57.

[14] Nietzsche, *The Will to Power* (trans. Walter Kaufmann and R.J. Hollingdale) Random House, New York, 1967, Book III, Section 539, p. 291. I learned of this

passage from Leonard Nash, *The Nature of the Natural Sciences*, Little and Brown, Boston, 1963, p. 362.
[15] *Conservation*, pp. 55–56.
[16] See R.G. Collingwood, *The Idea of Nature*, Clarendon Press, Oxford, 1945.
[17] For further discussion of this, see R.S. Cohen, 'Causation in History' in *Logic, Physical Reality, History* (ed. by Wolfgang Yourgrau), Plenum Press, New York, 1970).
[18] *La Connaissance et l'Erreur* (trans. by Marcel Dufour from the 2nd German edition of *Erkenntnis und Irrtum*), Flammarion, Paris, 1908, p. 368. (Hereafter cited as *Erreur*.)
[19] *Popular*, p. 193.
[20] *Ibid.*, p. 153.
[21] *Conservation*, p. 54.
[22] See the discussion in Hans Reichenbach, *Elements of Symbolic Logic*, Macmillan, New York, 1947, and in his *Theory of Probability*, 2nd ed. (trans. E.H. Hutten and Maria Reichenbach), University of California Press, Berkeley and Los Angeles, 1949.
[23] *Die Prinzipien der Wärmelehre*, 2nd ed. Johann Ambrosius Barth, Leipzig, 1900, p. 430, as translated by Stephen Brush. (Hereafter cited as *Wärmelehre*.)
[24] The critic was Hans Reichenbach in 'Ziele und Wege der physikalischen Erkenntnis', in *Handbuch der Physik*, Springer, Berlin, 1929, vol. IV, especially Section 6, 'Das Realitätsproblem', pp. 16–24.
[25] See K.R. Popper (as in reference 11, above); Duhem's instrumentalism was of partial meaning, or as we say today, of open texture; Mach was followed by Moritz Schlick and Stephen Toulmin.
[26] See the authoritative discussion of this in K.D. Heller, *Ernst Mach: Wegbereiter der Modernen Physik*, Springer, Vienna and New York, 1964, pp. 133–134.
[27] *Conservation*, p. 88. Compare the views of Joseph Agassi, 'The Confusion between Science and Technology in the Standard Philosophies of Science', *Technology and Culture* 7 (Summer, 1966) 348–366.
[28] Compare the more favorable opinion of Marx W. Wartofsky, 'Metaphysics as Heuristic for Science', in *Boston Studies in the Philosophy of Science*, III, (ed. by R.S. Cohen and M.W. Wartofsky), D. Reidel, Dordrecht, Holland, and Humanities Press, New York, 1967.
[29] *Popular*, p. 232.
[30] *Ibid.* See the discussion 'Le Darwinisme et l'idée d'adaptation', in Robert Bouvier, *La Pensée d'Ernst Mach*, Libraire au Velin D'Or, Paris, 1923, pp. 94–106.
[31] George Santayana, *Scepticism and Animal Faith*, Scribner, New York, 1923.
[32] Mach's discussion of evolutionary competition of theories as of species appears, e.g. in 'The Velocity of Light', in *Popular*, pp. 48–65.
[33] *Popular*, p. 232.
[34] *Erreur*, pp. 368–369, as cited in Bouvier (reference 30, above), p. 326.
[35] On the speculative character of the idea of the unity of science, see Joseph Agassi, 'Unity and Diversity in Science', in *Boston Studies in the Philosophy of Science*, IV, D. Reidel, Dordrecht, Holland, and Humanities Press, New York, 1969, p. 463.
[36] See the discussion of Paul Oppenheim and Hilary Putnam, 'Unity of Science as a Working Hypothesis', in *Minnesota Studies in the Philosophy of Science*, II (ed. by Herbert Feigl *et al.*), University of Minnesota Press, Minneapolis, 1958, pp. 3–36.
[37] Richard von Mises, *Positivism: A Study in Human Understanding*, Harvard University Press, Cambridge, Mass., 1951, Ch. 7: 'Mach's Elements', and Ch. 8: 'Protocol Sentences'.

[38] Ralph Barton Perry, *Present Philosophical Tendencies*, Longmans and Green, New York, 1912, pp. 78–79.
[39] *Analysis*, p. 60.
[40] *Analysis*, pp. 340–341; the citations in the entire paragraph derive from this passage.
[41] Sigmund Freud to Wilhelm Fliess, June 12, 1900, in *The Origins of Psychoanalysis: Letters to Wilhelm Fliess, Drafts and Notes: 1887–1902* (ed. by Marie Bonaparte *et al.*), Basic Books, New York, 1954, p. 322. Compare the somewhat different analysis of Thomas Szasz, 'Mach and Psychoanalysis', *J. Nervous and Mental Disease* **130** (January, 1960) 6–15.
[42] There is a substantial literature on the unity of science movement which was an offshoot of the Ernst Mach Verein (The Vienna Circle around Moritz Schlick). Otto Neurath was the principal organizer and spokesman. See, e.g., his 'Einheit der Wissenschaft als Aufgabe', *Erkenntnis* **5** (1935) 16–22; 'Einzelwissenschaften, Einheitswissenschaft, Pseudorationalismus', in *Actes du Congrès International de Philosophie Scientifique*, I, Hermann, Paris, 1936; 'L'Encyclopédie comme "modèle"', *Revue de Synthese* **12** (October, 1936) 187–201; 'Unified Science and its Encyclopedia', *Philosophy of Science* **4** (April, 1937) 265–277; 'Unified Science as Encyclopedic Integration', in *International Encyclopedia of Unified Science*, I, nr. 1 University of Chicago Press, Chicago, 1938, pp. 1–27.
[43] *Mechanics*, p. 560. I have modified the text according to the translation of von Mises in his *Positivism* (reference 37, above), p. 201.
[44] Husserl's discussion appears in his *Logische Untersuchungen* of 1900.
[45] *Mechanics*, pp. 593–595. All the citations in this paragraph come from these pages.
[46] Compare the critical expositions of Herbert Marcuse, 'On Science and Phenomenology' and Aron Gurwitsch, 'Comment on the Paper by H. Marcuse', in *Boston Studies in the Philosophy of Science*, II, Humanities Press, New York, 1965, pp. 279–290 and 291–306.
[47] *Erreur*, p. 13. See Emile Meyerson's discussion of this point, with particular reference to Poincaré's even stronger opinion, in *Identity and Reality* (trans. Kate Loewenberg), Macmillan, New York, and Allen and Unwin, London, 1930, p. 42.
[48] Sandor Ferenczi, *Final Contributions to the Problems and Methods of Psychoanalysis* Hogarth Press, London, 1955, p. 192.
[49] *Kultur und Mechanik*, W. Spemann, Stuttgart, 1915, as translated by Jane Suttie (see reference 51, below).
[50] The principal text is *Wärmelehre*, pp. 443–444. Two excerpts, in Jane Suttie's translation, follow:

"... the remarkable fact that an idea, so to say, *continues to exist and act without its being in consciousness*. In this connection the excellent observations of W. Robert concerning dreams (Seippel, Hamburg, 1886) might be illuminating. Robert has observed that those series of associations that are disturbed and interrupted by day are elaborated by night as dreams."

"It is quite probable, therefore, *that ideas once formed, even if they are no longer in consciousness, nevertheless maintain their existence*. This seems peculiarly the case if these ideas, on entering consciousness, have been prevented from liberating their associated ideas, movements, etc. In these circumstances they seem to act as a sort of *charge ... to a certain extent related to these are the phenomena which Breuer and Freud described recently in their book on hysteria*."

[51] Sandor Ferenczi, 'Concerning the Psychogenesis of Mechanism', *Imago* **5** (1919) 394,

and 'Supplement to "The Psychogenesis of Mechanism"', *Imago* **6** (1920) 384, both as translated by Jane Suttie and others and reprinted in *Further Contributions to the Theory and Technique of Psycho-Analysis*, Hogarth Press, London, 1926, pp. 383–396.
[52] See Freud's treatment of the rational syndrome of order, parsimony and obstinacy in *Collected Papers*, II, Hogarth Press, London, 1950, pp. 45–50.
[53] *Popular*, pp. 233–235.
[54] *Analysis*, p. 361. The second quoted sentence is modified in accord with Cecily Hastings (see reference 55, below).
[55] Erwin Schrödinger, *My View of the World* (trans. Cecily Hastings), Cambridge University Press, Cambridge, 1964, p. 17.
[56] *Ibid.*, p. 37.
[57] See the unsurpassed exposition in Bertrand Russell, *Mysticism and Logic*, Longmans, Green, London, 1918, pp. 1–32.
[58] For a detailed discussion, see my discussion of Lenin's book *Materialism and Empirio-Criticism* in Part IV 'The Dangers of Subjectivism' of the essay cited in reference 4 above.
[59] A useful, though opinionated survey, is to be found in Thomas G. Masaryk, *The Spirit of Russia: Studies in History, Literature and Philosophy* (trans. Eden and Cedar Paul), 2nd ed., with additions by Jan Slavik, Allen and Unwin, London, and Macmillan, New York, 1955, esp. vol. II, Ch. 15 on Lavrov and Mihailovski, pp. 115–190.
[60] See the account of Bogdanov by Gustav A. Wetter, 'Bogdanov's Empirio-Monism' in Wetter's *Dialectical Materialism: A Historical and Systematic Survey of Philosophy in the Soviet Union* (trans. Peter Heath), Routledge and Kegan Paul, London, 1958, Part One, Ch. IV: 4, pp. 92–99; also the article on him by George L. Kline in *Encyclopedia of Philosophy* (ed. by Paul Edwards) Macmillan and Free Press, New York, and Collier-Macmillan, London, 1967, vol. I, p. 331; further, the authorized current Soviet account (1959) as available in *Geschichte der Philosophie*, Deutscher Verlag der Wissenschaften, Berlin, 1961, III, pp. 318–328; and the excerpt 'Matter as Thing-in-Itself' (trans. George L. Kline), in *Russian Philosophy* (ed. by James M. Edie *et al.*), Quadrangle Books, Chicago, 1965), III, pp. 393–404. See also the detailed monograph by Dietrich Grille, *Lenins Rivale: Bogdanov und seine Philosophie* (Abhandlungen des Bundesinstituts für estwissenschaftliche und internationale Studien, Bd. XII), Verlag Wissenschaft und Politik, Cologne, 1966, which contains a complete bibliography of Bogdanov's writings, a German translation of Bogdanov's brief autobiography of 1926, and a full account of Machism in Czarist Russia.
A long section from the introduction to the third volume of Bogdanov's *Empiriomonism*, pp. xxii–xxxix, Petersburg, 1906, has recently been published in French translation under the title 'La voie de l'empiriomonisme', together with a useful explanatory essay by the translator, A. Yassour: see *Cahiers de l'Institut de science économique appliquée* S, 10, No. 176, (Paris, August 1966) pp. 139–172.
[61] Philipp Frank, among the Machists, wrote of these parallels on several occasions, and so did Otto Neurath. Neurath is, in a sense quite different from Bogdanov's outlook, also a Marxist-Machist. For Frank's view of the parallels, see *Das Kausalgesetz und seine Grenzen*, Springer, Vienna, 1932, Ch. X, Sect. 13–14, and also his *Modern Science and Its Philosophy*, Harvard University Press, Cambridge, Mass., 1949. Ch. 11, 'Logical Empiricism and the Philosophy of the Soviet Union', pp. 198–206. For Neurath, see *Lebensgestaltung und Klassenkampf*, E. Laub, Berlin, 1928, 'Weltanschauung und Marxismus', *Der Kampf* **24** (Vienna, 1931) 447–451, and especially *Empirische Soziologie: Die wissenschaftliche Gehalt der Geschichte und Nationalökono-*

mie, Springer, Vienna, 1931. The relevant sections of the last work will appear in English as part of a larger anthology of Neurath's work: *Empirical Sociology: The Life and Work of Otto Neurath* (ed. by Marie Neurath and R.S. Cohen), D. Reidel, Dordrecht, 1970.

62 See Herbert Marcuse, 'Philosophie und kritische Theorie', *Z. f. Sozialforschung* **6** (1937) 632, as reprinted in his *Kultur und Gesellschaft*, I, Suhrkamp, Frankfurt-am-Main, 1965, pp. 102–127, and also Max Horkheimer, 'Der neueste Angriff auf die Metaphysik', *Z. f. Sozialforschung* **6** (1937) 4–51. The same tendency appears among more partisan Marxists too, e.g. Leszek Kolakowski's still untranslated work on 'the individual and the infinite', *Jednostka i nieskończoność: Wolność i antynomie wolności w filozofii Spinozy*, PWN, Warsaw, 1958 also see the variety of expositions of social criticism in the collective work, *Socialist Humanism* (ed. by Erich Fromm), Doubleday, New York, 1965.

63 Much of the popular writings of the Vienna Circle was devoted to this theme of anti-metaphysical enlightenment. For a brief account, see the original statement by Neurath, Hans Hahn and Rudolf Carnap: *The Vienna Circle: Scientific Conception of the World* (transl. from the Vienna edition of 1929), in *Empirical Sociology* (reference 61, above).

64 See the long excerpt from Gyorgy Lukacs, *Marxismus und Existentialismus*, translated as 'Existentialism' in *Philosophy for the Future: The Quest of Modern Materialism* (ed. by R.W. Sellars *et al.*), Macmillan, New York, 1949, pp. 571–590, esp. 572.

65 In his influential essay of 1937, Mao Tse-Tung reaffirmed the Marxist critique of mere sense-perception: "Our practice proves that things perceived cannot be readily understood by us, and that only things understood can be more profoundly perceived. Perception only solves the problem of phenomena; reason alone solves the problem of essence", from 'On Practice', in *Selected Works*, I, Lawrence and Wishart, London, and International Publishers, New York, 1937, p. 286.

66 A clear survey is given in Adam Schaff's work *Zu einigen Fragen der Marxistischen Theorie der Wahrheit*, German trans. from Polish (ed. by G. Klaus), Dietz, Berlin, 1954.

GERALD J. HOLTON

MACH, EINSTEIN, AND THE SEARCH FOR REALITY

In the history of ideas of our century, there is a chapter that might be entitled 'The Philosophical Pilgrimage of Albert Einstein', a pilgrimage from a philosophy of science in which sensationism and empiricism were at the center, to one in which the basis was a rational realism. This essay,* a portion of a more extensive study[1], is concerned with Einstein's gradual philosophical reorientation, particularly as it has become discernible during the work on his largely unpublished scientific correspondence.[2]

The earliest known letter by Einstein takes us right into the middle of the case. It is dated 19 March 1901 and addressed to Wilhelm Ostwald.[3] The immediate cause for Einstein's letter was his failure to receive an assistantship at the school where he had recently finished his formal studies, the Polytechnic Institute in Zürich; he now turned to Ostwald to ask for a position at his laboratory, partly in the hope of receiving "the opportunity for further education". Einstein included a copy of his first publication, 'Folgerungen aus den Capillaritätserscheinungen', *Ann. Physik*, **4** (1901), 513, which he said had been inspired (*angeregt*) by Ostwald's work; indeed, Ostwald's *Allgemeine Chemie* is the first book mentioned in all of Einstein's published work.

Not having received an answer, Einstein wrote again to Ostwald on 3 April 1901. On 13 April 1901 his father, Hermann Einstein, sent Ostwald a moving appeal, evidently without his son's knowledge. Hermann Einstein reported that his son esteems Ostwald "most highly among all scholars currently active in physics".[4]

The choice of Ostwald was significant. He was, of course, not only one of the foremost chemists, but also an active 'philosopher-scientist' during the 1890's and 1900's, a time of turmoil in the physical sciences as well as in the philosophy of science. The opponents of kinetic, mechanical, or materialistic views of natural phenomena were vociferous. They objected to atomic theory and gained great strength from the victories of thermodynamics, a field in which no knowledge or assumption was needed concerning the detailed nature of material substances (for example, for an understanding of heat engines).

Ostwald was a major critic of the mechanical interpretation of physical phenomena, as were Helm, Stallo, and Mach. Their form of positivism – as against the sophisticated logical positivism developed later in Carnap and Ayer's work – provided an epistemology for the new phenomenologically based science of correlated observations, linking energetics and sensationism. In the second (1893) edition of his influential textbook on chemistry, Ostwald had given up the mechanical treatment of his first edition for Helm's 'energetic' one. 'Hypothetical' quantities such as atomic entities were to be omitted; instead, these authors claimed they were satisfied, as Merz wrote around 1904, with "measuring such quantities as are presented directly in observation, such as energy, mass, pressure, volume, temperature, heat, electrical potential, etc., without reducing them to imaginary mechanisms or kinetic quantities". They condemned such conceptions as the ether, with properties not accessible to direct observation, and they issued a call "to consider anew the ultimate principles of all physical reasoning, notably the scope and validity of the Newtonian laws of motion and of the conceptions of force and action, of absolute and relative motion".[5]

All these iconoclastic demands – except anti-atomism – must have been congenial to the young Einstein who, according to his colleague Joseph Sauter, was fond of calling himself 'a heretic'.[6] Thus, we may well suspect that Einstein felt sympathetic to Ostwald who denied in the *Allgemeine Chemie*[7] that "the assumption of that medium, the ether, is unavoidable. To me it does not seem to be so. ... There is no need to inquire for a carrier of it when we find it anywhere. This enables us to look upon radiant energy as independently existing in space". It is a position quite consistent with that shown later in Einstein's papers of 1905 on photon theory and relativity theory.

In addition, it is worth noting that Einstein, in applying to Ostwald's laboratory, seemed to conceive of himself as an experimentalist. We know from many sources that in his student years in Zürich Einstein's earlier childhood interest in mathematics had slackened considerably. In the *Autobiographical Notes*[8], Einstein reported: "I really could have gotten a sound mathematical education. However, I worked most of the time in the physical laboratory, fascinated by the direct contact with experience"

(p. 15). To this, one of his few reliable biographers adds: "No one could stir him to visit the mathematical seminars. ... He did not yet see the possibility of seizing that formative power resident in mathematics, which later became the guide of his work. ... He wanted to proceed quite empirically, to suit his scientific feeling of the time. ... As a natural scientist, he was a pure empiricist." (Anton Reiser, *Albert Einstein*, New York 1930, pp. 51–52).

Ostwald's main philosophical ally was the prolific and versatile Austrian physicist and philosopher Ernst Mach (1838–1916), whose main work Einstein had read avidly in his student years and with whom he was destined to have later the encounters that form a main concern of this paper. Mach's major book, *The Science of Mechanics*[9], first published in 1883, is perhaps most widely known for its discussion of Newton's *Principia*, in particular for its devastating critique of what Mach called the "conceptual monstrosity of absolute space" (Preface, 7th Edition, 1912) – a conceptual monstrosity because it is "purely a thought-thing which cannot be pointed to in experience". Starting from his analysis of Newtonian presuppositions, Mach proceeded in his announced program of eliminating all metaphysical ideas from science. As Mach said quite bluntly in the preface to the first edition of *The Science of Mechanics*: "This work is not a text to drill theorems of mechanics. Rather, its intention is an enlightening one – or to put it still more plainly, an anti-metaphysical one."

It will be useful to review briefly the essential points of Mach's philosophy. Here we can benefit from a good, although virtually unknown, summary presented by his sympathetic follower, Moritz Schlick, in the essay 'Ernst Mach, Der Philosoph'.[10]

Mach was a physicist, physiologist, and also psychologist, and his philosophy ... arose from the wish to find a principal point of view to which he could hew in any research, one which he would not have to change when going from the field of physics to that of physiology or psychology. Such a firm point of view he reached by going back to that which is given before all scientific research: namely, the world of sensations. ... Since all our testimony concerning the so-called external world rely only on sensations, Mach held that we can and must take these sensations and complexes of sensations to be the sole contents [*Gegenstände*] of those testimonies, and, therefore, that there is no need to assume in addition an unknown reality hidden behind the sensations. With that, the existence *der Dinge an sich* is removed as an unjustified and unnecessary assumption. A body, a physical object, is nothing else than a complex, a more or less firm [we would say, invariant] pattern of sensations, i.e., of colors, sounds, sensations of heat and pressure, etc.

> There exists in this world nothing whatever other than sensations and their connections. In place of the word "sensations," Mach liked to use rather the more neutral word "elements." ... [As is particularly clear in Mach's book *Erkenntnis und Irrtum*] scientific knowledge of the world consists, according to Mach, in nothing else than the simplest possible description of the connections between the elements, and it has as its only aim the intellectual mastery of those facts by means of the least possible effort of thought. This aim is reached by means of a more and more complete "accommodation of the thoughts to one another." This is the formulation by Mach of his famous "principle of the economy of thought."[11]

The influence of Mach's point of view, particularly in the German-speaking countries, was enormous – on physics, on physiology, on psychology, and on the fields of the history and the philosophy of science[12] (not to mention Mach's profound effect on the young Lenin, Hofmannstal, Musil, among many others outside the sciences). Strangely neglected by recent scholarship – there is not even a major biography – Mach has in the last two or three years again become the subject of a number of promising studies. To be sure, Mach himself always liked to insist that he was beleaguered and neglected, and that he did not have, or wish to have, a philosophical system; yet his philosophical ideas and attitudes had become so widely a part of the intellectual equipment of the period from the 1880's on that Einstein was quite right in saying later that "even those who think of themselves as Mach's opponents hardly know how much of Mach's views they have, as it were, imbided with their mother's milk".[13]

The problems of physics themselves at that time helped to reinforce the appeal of the new philosophical attitude urged by Mach. The great program of 19th-century physics, the reconciliation of the notions of ether, matter, and electricity by means of mechanistic pictures and hypotheses, had led to enormities – for example, Larmor's proposal that the electron is a permanent but movable state of twist or strain in the ether, forming discontinuous particles of electricity and possibly of all ponderable matter. To many of the younger physicists of the time, attacking the problems of physics with conceptions inherited from classical nineteenth century, physics did not seem to lead anywhere. And here Mach's iconoclasm and incisive critical courage, if not the details of his philosophy, made a strong impression on his readers.

I. MACH'S EARLY INFLUENCE ON EINSTEIN

As the correspondence at the Einstein Archives at Princeton reveals, one of the young scientists deeply caught up in Mach's point of view was Michelange (Michele) Besso – Einstein's oldest and closest friend, fellow student, and colleague at the Patent Office in Bern, the only person to whom Einstein gave public credit for help (*manche wertvolle Anregung*) when he published his basic paper on relativity in 1905. It was Besso who introduced Einstein to Mach's work. In a letter of 8 April 1952 to Carl Seelig, Einstein wrote: "My attention was drawn to Ernst Mach's *Science of Mechanics* by my friend Besso while a student, around the year 1897. The book exerted a deep and persisting impression upon me ..., owing to its physical orientation toward fundamental concepts and fundamental laws." As Einstein noted in his *Autobiographical Notes*[8] written in 1946, Ernst Mach's *The Science of Mechanics* "shook this dogmatic faith" in "mechanics as the final basis of all physical thinking. ... This book exercised a profound influence upon me in this regard while I was a student. I see Mach's greatness in his incorruptible skepticism and independence; in my younger years, however, Mach's epistemological position also influenced me very greatly" (p. 21).

As the long correspondence between those old friends shows, Besso remained a loyal Machist to the end. Thus, writing to Einstein on 8 December 1947, he still said: "As far as the history of science is concerned, it appears to me that Mach stands at the center of the development of the last 50 or 70 years." Is it not true, Besso also asked, "that this introduction [to Mach] fell into a phase of development of the young physicist [Einstein] when the Machist style of thinking pointed decisively at observables – perhaps even, indirectly, to clocks and meter sticks?"

Turning now to Einstein's crucial first paper on relativity in 1905, we can discern in it influences of many, partly contradictory, points of view – not surprising in a work of such originality by a young contributor. Elsewhere I have examined the effect – or lack of effect – on that paper of three contemporary physicists: H. A. Lorentz[14], Henri Poincaré[15], and August Föppl.[1] Here we may ask in what sense and to what extent Einstein's initial relativity paper of 1905 was imbued with the style of thinking associated with Ernst Mach and his followers – apart from the characteristics of clarity and independence, the two traits in Mach which Einstein always praised most.

In brief, the answer is that the Machist component – a strong component, even if not the whole story – shows up prominently in two related respects: first, by Einstein's insistence from the beginning of his relativity paper that the fundamental problems of physics cannot be understood until an epistemological analysis is carried out, particularly so with respect to the meaning of the conceptions of space and time[16]; and second, by Einstein's identification of reality with what is given by sensations, the 'events', rather than putting reality on a plane beyond or behind sense experience.

From the outset, the instrumentalist, and hence sensationist, views of measurement and of the concepts of space and time are strikingly evident. The key concept in the early part of the 1905 paper is introduced at the top of the third page in a straightforward way. Indeed, Leopold Infeld in his biography of Einstein called them "the simplest sentence[s] I have ever encountered in a scientific paper". Einstein wrote: "We have to take into account that all our judgments in which time plays a part are always judgments of *simultaneous events*. If for instance I say, 'that train arrived here at seven o'clock', I mean something like this: 'The pointing of the small hand of my watch to seven and the arrival of the train are simultaneous events.'"[17]

The basic concept introduced here, one that overlaps almost entirely Mach's basic 'elements', is Einstein's concept of *events* (*Ereignisse*) – a word that recurs in Einstein's paper about a dozen times immediately following this citation. Transposed into Minkowski's later formulation of relativity, Einstein's 'events' are the intersections of particular 'world lines', say that of the train and that of the clock. The time (*t coordinate*) of an event by itself has no operational meaning. As Einstein says: "The 'time' of an event is that which is given simultaneously with the event by a stationary clock located at the place of the event" (p. 894). We can say that just as the *time* of an event assumes meaning only when it connects with our consciousness through sense experience (that is, when it is subjected to measurement-in-principle by means of a clock present at the same place), so also is the *place*, or space co-ordinate, of an event meaningful only if it enters our sensory experience while being subjected to measurement-in-principle (that is, by means of meter sticks present on that occasion at the same time).[18]

This was the kind of operationalist message which, for most of his readers, overshadowed all other philosophical aspects in Einstein's paper. His work was enthusiastically embraced by the groups who saw themselves as philosophical heirs of Mach, the Vienna Circle of neopositivists and its predecessors and related followers,[19] providing a tremendous boost for the philosophy that had initially helped to nurture it. A typical response welcoming the relativity theory as "the victory over the metaphysics of absolutes in the conceptions of space and time ... a mighty impulse for the development of the philosophical point of view of our time", was extended by J. Petzoldt in the inaugural session of the *Gesellschaft für Positivistische Philosophie* in Berlin, 11 November 1912.[20] Michele Besso, who had heard the message from Einstein before anyone else, had exclaimed: "In the setting of Minkowski's spacetime frame work, it was now first possible to carry through the thought which the great mathematician, Bernhard Riemann, had grasped: 'The space-time framework itself is formed by the events in it.'"[21]

To be sure, re-reading Einstein's paper with the wisdom of hindsight, as we shall do presently, we can find in it also very different trends, warning of the possibility that 'reality' in the end is not going to be left identical with 'events'. There are premonitions that sensory experiences, in Einstein's later work, will not be regarded as the chief building blocks of the 'world', that the laws of physics themselves will be seen to be built into the event-world as the undergirding structure 'governing' the pattern of events.

Such precursors appear even earlier, in one of Einstein's early letters in the Archives. Addressed to his friend Marcel Grossmann, it is dated 14 April 1901, when Einstein believed he had found a connection between Newtonian forces and the forces of attraction between molecules: "It is a wonderful feeling to recognize the unity of a complex of appearances which, to direct sense experience, seem to be separate things." Already there is a hint here of the high value that will be placed on intuited unity and the limited role seen for evident sense experience.

But all this was not yet ready to come into full view, even to the author. Taking the early papers as a whole, and in the context of the physics of the day, we find that Einstein's philosophical pilgrimage did start on the historic ground of positivism. Moreover, Einstein thought so himself, and confessed as much in letters to Ernst Mach.

II. THE EINSTEIN–MACH LETTERS

In the history of recent science, the relation between Einstein and Mach is an important topic that has begun to interest a number of scholars. Indeed, it is a drama of which we can sketch here four stages: Einstein's early acceptance of the main features of Mach's doctrine; the Einstein–Mach correspondence and meeting; the revelation in 1921 of Mach's unexpected and vigorous attack on Einstein's relativity theory; and Einstein's own further development of a philosophy of knowledge in which he rejected many, if not all, of his earlier Machist beliefs.

Happily, the correspondence is preserved at least in part. A few letters have been found, all from Einstein to Mach. Those of concern here are part of an exchange between 1909 and 1913, and they testify to Einstein's deeply felt attraction to Mach's viewpoint, just at a time when the mighty Mach himself – forty years senior to the young Einstein whose work was just becoming widely known – had for his part embraced the relativity theory publicly by writing in the second (1909) edition of *Conservation of Energy*: "I subscribe, then, to the principle of relativity, which is also firmly upheld in my *Mechanics* and *Wärmelehre*."[22] In the first letter, Einstein writes from Berne on 9 August 1909. Having thanked Mach for sending him the book on the law of conservation of energy, he adds: "I know, of course, your main publications very well, of which I most admire your book on Mechanics. You have had such a strong influence upon the epistemological conceptions of the younger generation of physicists that even your opponents today, such as Planck, undoubtedly would have been called Mach followers by physicists of the kind that was typical a few decades ago."

It will be important for our analysis to remember that Planck was Einstein's earliest patron in scientific circles. It was Planck who, in 1905, as editor of the *Annalen der Physik*, received Einstein's first relativity paper and thereupon held a review seminar on the paper in Berlin. Planck defended Einstein's work on relativity in public meetings from the beginning, and by 1913 had succeeded in persuading his German colleagues to invite Einstein to the Kaiser-Wilhelm-Gesellschaft in Berlin. With a polemical essay 'Against the New Energetics' in 1896, he had made clear his position, and by 1909 Planck was one of the few opponents of Mach, and scientifically the most prominent one. He had just written a famous attack, *Die Einheit des physikalischen Weltbildes*. Far from accepting Mach's view that, as he put it, "Nothing is real except the perceptions, and all natural science is ultimately an economic adaptation of our ideas to our perceptions", Planck held to the entirely antithetical position that a basic aim of science is "the finding of a *fixed* world picture independent of the variation of time and people", or, more generally, "the complete liberation of the physical picture from the individuality of the separate intellects".[23] At least by implication in Einstein's remarks to Mach, he dissociated himself from allegiance to Planck's view. It may also not be irrelevant that just at that time Einstein, who since 1906 had been objecting to inconsistencies in Planck's quantum theory, was preparing his first major invited paper before a scientific congress, the eighty-first meeting of the Naturforscherversammlung, announced for September, 1909, in Salzburg. Einstein's paper called for a revision of Maxwell's theory to accommodate the probabilistic character of the emission of photons – none of which Planck could accept – and concluded: "To accept Planck's theory means, in my view, to throw out the bases of our [1905] theory of radiation."

Mach's reply to Einstein's first letter is now lost, but it must have come quickly, because eight days later Einstein sends an acknowledgment:

Berne, 17 August 1909. Your friendly letter gave me enormous pleasure. ... I am very glad that you are pleased with the relativity theory. ... Thanking you again for your friendly letter, I remain, your student [indeed: *Ihr Sie verehrender Schüler*], A. Einstein.

Einstein's next letter was written as physics professor in Prague, where Mach before him had been for twenty-eight years. The post had been offered to Einstein on the basis of recommendations of a faction (Lampa, Pick) who regarded themselves as faithful disciples of Mach. The letter was sent out about New Year's 1911–12, perhaps just before or after Einstein's sole (and, according to P. Frank's account, in *Einstein, His Life and Times*, not very successful) visit to Mach, and after the first progress toward the general relativity theory.

... I can't quite understand how Planck has so little understanding for your efforts. His stand to my [general relativity] theory is also one of refusal. But I can't take it amiss; so far, that one single epistemological argument is the only thing which I can bring forward in favor of my theory.

Here, Einstein is referring delicately to the Mach Principle, which he had been putting at the center of the developing theory.[24] Mach responded by sending Einstein a copy of one of his books, probably the *Analysis of Sensations*.

In the last of these letters to Mach (who was now seventy-five years old, and for some years had been paralyzed), Einstein writes from Zürich on 25 June 1913:

Recently you have probably received my new publication on Relativity and Gravitation which I have at last finished after unending labor and painful doubt. [This must have been the "Entwurf einer verallgemeinerten Relativitätstheorie und einer Theorie der Gravitation", written with Marcel Grossmann.[25]] Next year at the solar eclipse it will turn out whether the light rays are bent by the sun, in other words whether the basic and fundamental assumption of the equivalence of the acceleration of the reference frame and of the gravitational field really holds. If so, then your inspired investigations into the foundations of mechanics – despite Planck's unjust criticism – will receive a splendid confirmation. For it is a necessary consequence that inertia has its origin in a kind of mutual interaction of bodies, fully in the sense of your critique of Newton's bucket experiment.[26]

III. THE PATHS DIVERGE

The significant correspondence stops here, but Einstein's public and private avowals of his adherence to Mach's ideas continue for several years more. For example, there is his well-known, moving eulogy of Mach, published in 1916.[13] In August, 1918, Einstein writes to Besso quite sternly about an apparent – and quite temporary – lapse in Besso's positivistic epistemology; it is an interesting letter, worth citing in full:

28 August 1918.

Dear Michele:

In your last letter I find, on re-reading, something which makes me angry: That speculation has proved itself to be superior to empiricism. You are thinking here about the development of relativity theory. However, I find that this development teaches something else, that it is practically the opposite, namely that a theory which wishes to deserve trust must be built upon generalizable facts.

Old examples: Chief postulates of thermodynamics [based] on impossibility of perpetuum mobile. Mechanics [based] on grasped [*ertasteten*] law of inertia. Kinetic gas theory [based] on equivalence of heat and mechanical energy (also historically). Special Relativity on the constancy of light velocity and Maxwell's equation for the vacuum, which in turn rest on empirical foundations. Relativity with respect to uniform [?] translation is a *fact of experience.*

General Reality: *Equivalence of inertial and gravitational mass.* Never has a truly useful and deep-going theory really been found purely speculatively. The nearest case is Maxwell's hypothesis concerning displacement current; there the problem was to do justice to the fact of light propagation. ... With cordial greetings, your Albert. [Emphases in original]

Careful reading of this letter shows us that already here there is evidence of divergence between the conception of 'fact' as understood by Einstein and 'fact' as understood by a true Machist. The impossibility of the *perpetuum mobile*, the first law of Newton, the constancy of light velocity, the validity of Maxwell's equations, the equivalence of inertial and gravitational mass – none of these would have been called 'facts of experience' by Machists. Indeed, Mach might have insisted that – to use one of his favorite battle words – it is evidence of 'dogmatism' not to regard all these conceptual constructs as continually in need of probing re-examination; thus, Mach had written:[27]

... for me, matter, time and space are still *problems*, to which, incidentally, the physicists (Lorentz, Einstein, Minkowski) are also slowly approaching.

Similar evidence of Einstein's gradual apostasy appears in a letter of 4 December 1919 to Paul Ehrenfest. Einstein writes:

I understand your difficulties with the development of relativity theory. They arise simply because you want to base the innovations of 1905 on epistemological grounds (non-existence of the stagnant ether) instead of empirical grounds (equivalence of all inertial systems with respect to light).

Mach would have applauded Einstein's life-long suspicion of formal epistemological systems, but how strange would he have found this use of the word *empirical* to characterize the hypothesis of the equivalence of all inertial systems with respect to light! What we see forming slowly here is Einstein's view that the fundamental role played by experience in the construction of fundamental physical theory is, after all, not through the 'atom' of experience, not through the individual sensation or the protocol sentence, but through some creative digest or synthesis of '*die gesammten Erfahrungstatsachen*', the *totality* of physical experience.[28] But all this was still hidden. Until Mach's death, and for several years after, Einstein considered and declared himself a disciple of Mach.

In the meantime, however, unknown to Einstein and everyone else, a time bomb had been ticking away. Set in 1913, it went off in 1921, five years after Mach's death, when Mach's *The Principles of Physical Optics* was published at last. Mach's preface was dated July, 1913 – perhaps a few days or, at most, a few weeks after Mach had received Einstein's last, enthusiastic letter and the article on general relativity theory. In a well-known passage in the Preface (but one usually found in an inaccurate translation), Mach had written:

I am compelled, in what may be my last opportunity, to cancel my views [*Anschauungen*] of the relativity theory.

I gather from the publications which have reached me, and especially from my correspondence, that I am gradually becoming regarded as the forerunner of relativity. I am able even now to picture approximately what new expositions and interpretations many of the ideas expressed in my book on Mechanics will receive in the future from this point of view. It was to be expected that philosophers and physicists should carry on a crusade against me, for, as I have repeatedly observed, I was merely an unprejudiced rambler endowed with original ideas, in varied fields of knowledge. I must, however, as assuredly disclaim to be a forerunner of the relativists as I personally reject the atomistic doctrine of the present-day school, or church. The reason why, and the extent to which, I reject [*ablehne*] the present-day relativity theory, which I find to be growing more and more dogmatical, together with the particular reasons which have led me to such a view – considerations based on the physiology of the senses, epistemological doubts, and above all the insight resulting from my experiments – must remain to be treated in the sequel [a sequel which was never published].

Certainly, Einstein was deeply disappointed by this belated disclosure of Mach's sudden dismissal of the relativity theory. Some months later, during a lecture on 6 April 1922 in Paris, in a discussion with the anti-Machist philosopher Emile Meyerson, Einstein allowed in a widely reported remark that Mach was "*un bon mécanicien*", but a "*déplorable philosophe*".[29]

We can well understand that Mach's rejection was at heart very painful, the more so as it was somehow Einstein's tragic fate to have the contribution he most cared about rejected by the very men whose approval and understanding he would have most gladly had – a situation not unknown in the history of science. In addition to Mach, the list includes these four: *H. Poincaré*, who, to his death in 1912, only once deigned to mention Einstein's name in print, and then only to register an objection; *H. A. Lorentz*, who gave Einstein personally every possible encouragement – short of fully accepting the theory of relativity for himself; *Planck*, whose support of the special theory of relativity was unstinting, but who resisted Einstein's ideas on general relativity and the early quantum theory of radiation; and *A. A. Michelson*, who to the end of his days did not believe in relativity theory, and once said to Einstein that he was sorry that his own work may have helped to start this "monster".[30]

Soon Einstein's generosity again took the upper hand and resulted, from then to the end of his life, in many further personal testimonies to Mach's earlier influence.[31] A detailed analysis was provided in Einstein's letter of 8 January 1948 to Besso:

As far as Mach is concerned, I wish to differentiate between Mach's influence in general and his influence on me. ... Particularly in the *Mechanics* and the *Wärmelehre* he tried to show how conceptions arose out of experience. He took convincingly the position that these conceptions, even the most fundamental ones, obtained their warrant only out of empirical knowledge, that they are in no way logically necessary. ...

I see his weakness in this, that he more or less believed science to consist in a mere ordering of empirical material; that is to say, he did not recognize the freely constructive element in formation of concepts. In a way he thought that theories arise through *discoveries* and not through *inventions*. He even went so far that he regarded "sensations" not only as material which has to be investigated, but, as it were, as the building blocks of the real world; thereby, he believed, he could overcome the difference between psychology and physics. If he had drawn the full consequences, he would have had to reject not only atomism but also the idea of a physical reality.

Now, as far as Mach's influence on my own development is concerned, it certainly was great. I remember very well that you drew my attention to his *Mechanics* and *Wärmelehre* during my first years of study, and that both books made a great impression on me. The extent to which they influenced my own work is, to say the truth, not clear to me. As far as I am conscious of it, the immediate influence of Hume on me was greater. ... But, as I said, I am not able to analyze that which lies anchored in unconscious thought. It is interesting, by the way, that Mach rejected the special relativity theory passionately (he did not live to see the general relativity theory [in the developed form]). The theory was, for him, inadmissibly speculative. He did not know that this speculative character belongs also to Newton's mechanics, and to every theory which thought is capable of. There exists only a gradual difference between theories, insofar as the chains of thought from fundamental concepts to empirically verifiable conclusions are of different lengths and complications.[32]

IV. ANTIPOSITIVISTIC COMPONENT OF EINSTEIN'S WORK

Ernst Mach's harsh words in his 1913 Preface leave a tantalizing mystery. Ludwig Mach's destruction of his father's papers has so far made it impossible to find out more about the 'experiments' (possibly on the constancy of the velocity of light) at which Ernst Mach hinted. Since 1921, many speculations have been offered to explain Mach's remarks.[33] They all leave something to be desired. Yet, I believe it is not so difficult to reconstruct the main reasons why Mach ended up rejecting the relativity theory. To put it very simply, Mach had recognized more and more clearly, years before Einstein did so himself, that Einstein had indeed fallen away from the faith, had left behind him the confines of Machist empiriocriticism.

The list of evidences is long. Here only a few examples can be given, the first from the 1905 relativity paper itself: What had made it really work was that it contained and combined elements based on two entirely different philosophies of science – not merely the empiricist-operationist component, but the courageous initial postulation, in the second paragraph, of two thematic hypotheses (one on the constancy of light velocity and the other on the extension of the principle of relativity to all branches of physics), two postulates for which there was and can be no direct empirical confirmation.

For a long time, Einstein did not draw attention to this feature. In a lecture at King's College, London, in 1921, just before the posthumous publication of Mach's attack, Einstein still was protesting that the origin of relativity theory lay in the facts of direct experience:

> ... I am anxious to draw attention to the fact that this theory is not speculative in origin; it owes its invention entirely to the desire to make physical theory fit observed fact as well as possible. We have here no revolutionary act, but the natural continuation of a line that can be traced through centuries. The abandonment of certain notions connected with space, time, and motion, hitherto treated as fundamentals, must not be regarded as arbitrary, but only as conditioned by observed facts.[34]

By June, 1933, however, when Einstein returned to England to give the Herbert Spencer Lecture at Oxford entitled 'On the Method of Theoretical Physics', the more complex epistemology that was in fact inherent in his work from the beginning had begun to be expressed. He opened this lecture with the significant sentence: "If you want to find out anything from the theoretical physicists about the methods they use, I advise you to stick closely to one principle: Don't listen to their words, fix your attention on their deeds." He went on to divide the tasks of experience and reason in a very different way from that advocated in his earlier visit to England:

> We are concerned with the eternal antithesis between the two inseparable components of our knowledge, the empirical and the rational. ... The structure of the system is the work of reason; the empirical contents and their mutual relations must find their representation in the conclusions of the theory. In the possibility of such a representation lie the sole value and justification of the whole system, and especially the concepts and fundamental principles which underlie it. Apart from that, these latter are free inventions of the human intellect, which cannot be justified either by the nature of that intellect or in any other fashion *a priori*.

In the summary of this section, he draws attention to the "purely fictitious character of the fundamentals of scientific theory". It is this penetrating insight which Mach must have smelled out much earlier and dismissed as "dogmatism".

Indeed Einstein, in his 1933 Spencer Lecture – widely read, as were and still are so many of his essays – castigates the old view that "the fundamental concepts and postulates of physics were not in the logical sense inventions of the human mind but could be deduced from experience by 'abstraction' – that is to say, by logical means. A clear recognition of the erroneousness of this notion really only came with the general theory of relativity".

Einstein ends this discussion with the enunciation of his current credo, so far from that he had expressed earlier:

> Nature is the realization of the simplest conceivable mathematical ideas. I am convinced that we can discover, by means of purely mathematical constructions, those concepts and those lawful connections between them which furnish the key to the understanding of natural phenomena. Experience may suggest the appropriate mathematical concepts, but they most certainly cannot be deduced from it. Experience remains, of course, the sole criterion of physical utility of a mathematical construction. But the creative principle resides in mathematics. In a certain sense, therefore, I hold it true that pure thought can grasp reality, as the ancients dreamed.[35]

Technically, Einstein was now at – or rather just past – the midstage of his pilgrimage. He had long ago abandoned his youthful allegiance to a primitive phenomenalism that Mach would have commended. In the first of the two passages just cited and others like it, he had gone on to a more refined form of phenomenalism which many of the logical positivists could still accept. He has, however, gone beyond it in the second passage, turning toward interests that we shall see later to have matured into clearly metaphysical conceptions.

Later, Einstein himself stressed the key role of what we have called thematic rather than phenomenic elements[36] – and thereby he fixed the early date at which, in retrospect, he found this need to arise in his earliest work. Thus he wrote in his *Autobiographical Notes* of 1946 that "shortly after 1900 ... I despaired of the possibility of discovering the true laws by means of constructive efforts based on known facts. The longer and the more despairingly I tried, the more I came to the conviction that only the discovery of *a universal formal principle* could lead us to assured results."[37]

Another example of evidence of the undercurrent of disengagement from a Machist position is an early one: It comes from Einstein's article on relativity in the 1907 *Jahrbuch der Radioactivität und Elektronik* (Vol. 4, No. 4), where Einstein responds, after a year's silence to W. Kaufmann's paper in *Ann. Physik* **19**, 1906. That paper had been the first publication in the *Annalen* to mention Einstein's work on the relativity theory, published there the previous year. Coming from the eminent experimental physicist Kaufmann, it had been most significant that this very first discussion was announced as a categorical, experimental disproof of Einstein's theory. Kaufmann had begun his attack with the devastating summary:

I anticipate right here the general result of the measurements to be described in the following: *the measurement results are not compatible with the Lorentz-Einsteinian fundamental assumption.*[38]

Einstein could not have known that Kaufmann's equipment was inadequate. Indeed, it took ten years for this to be fully realized, through the work of Guye and Lavanchy in 1916. So in his discussion of 1907, Einstein had to acknowledge that there seemed to be small but significant differences between Kaufmann's results and Einstein's predictions. He agreed that Kaufmann's calculations seemed to be free of error, but "whether there is an unsuspected systematic error or whether the foundations of relativity theory do not correspond with the facts one will be able to decide with certainty only if a great variety of observational material is at hand".

Despite this prophetic remark, Einstein does not rest his case on it. On the contrary, he has a very different, and what for his time and situation must have been a very daring, point to make: He acknowledges that the theories of electron motion given earlier by Abraham and by Bucherer do give predictions considerably closer to the experimental results of Kaufmann. But Einstein refuses to let the "facts" decide the matter: "In my opinion both theories have a rather small probability, because their fundamental assumptions concerning the mass of moving electrons are not explainable in terms of theoretical systems which embrace a greater complex of phenomena."[39]

This is the characteristic position – the crucial difference between Einstein and those who make the correspondence with experimental fact the chief deciding factor for or against a theory: Even though the "experimental facts" at that time very clearly seemed to favor the theory of his opponents rather than his own, he finds the *ad hoc* character of their theories more significant and objectionable than an apparent disagreement between his theory and their "facts".[40]

So already in this 1907 article – which, incidentally, Einstein mentions in his postcard of 17 August 1909 to Ernst Mach, with a remark regretting that he has no more reprints for distribution – we have explicit evidence of a hardening of Einstein against the epistemological priority of experiment, not to speak of sensory experience. In the years that followed, Einstein more and more openly put the consistency of a simple and convincing theory or of a thematic conception higher in importance than the latest news from the laboratory – and again and again he turned out to be right.

Thus, only a few months after Einstein had written in his fourth letter to Mach that the solar eclipse experiment will decide "whether the basic and fundamental assumption of the equivalence of the acceleration of the reference frame and of the gravitational field really holds", Einstein writes to Besso in a very different vein (in March 1914), before the first, ill-fated eclipse expedition was scheduled to test the conclusions of the preliminary version of the general relativity theory: "Now I am fully satisfied, and I do not doubt any more the correctness of the whole system, may the observation of the eclipse succeed or not. The sense of the thing [*die Vernunft der Sache*] is too evident." And later, commenting on the fact that there remains up to 10% discrepancy between the measured deviation of light owing to the sun's field and the calculated effect based on the general relativity theory: "For the expert, this thing is not particularly important, because the main significance of the theory does not lie in the verification of little effects, but rather in the great simplification of the theoretical basis of physics as a whole".[41] Or again, in Einstein's 'Notes on the Origin of the General Theory of Relativity',[42] he reports that he "was in the highest degree amazed" by the existence of the equivalence between inertial and gravitational mass, but that he "had no serious doubts about its strict validity, even without knowing the results of the admirable experiment of Eötvös".

The same point is made again in a revealing account given by Einstein's student, Ilse Rosenthal-Schneider. In a manuscript 'Reminiscences of Conversation with Einstein', dated 23 July 1957, she reports:

> Once when I was with Einstein in order to read with him a work that contained many objections against his theory ... he suddenly interrupted the discussion of the book, reached for a telegram that was lying on the windowsill, and handed it to me with the words, "Here, this will perhaps interest you". It was Eddington's cable with the results of measurement of the eclipse expedition [1919]. When I was giving expression to my joy that the results coincided with his calculations, he said quite unmoved, "But I knew that the theory is correct"; and when I asked, what if there had been no confirmation of his prediction, he countered: "Then I would have been sorry for the dear Lord – the theory *is* correct."[43]

V. MINKOWSKI'S 'WORLD' AND THE WORLD OF SENSATIONS

The third major point at which Mach, if not Einstein himself, must have seen that their paths were diverging is the development of relativity theory into the geometry of the four-dimensional space-time continuum, begun in 1907 by the mathematician H. Minkowski (who, incidentally, had had Einstein as a student in Zürich). Indeed, it was through Minkowski's semipopular lecture, 'Space and Time', on 21 September 1908 at the eightieth meeting of the Naturforscherversammlung[44], that a number of scientists first became intrigued with relativity theory. We have several indications that Mach, too, was both interested in and concerned about the introduction of four-dimensional geometry into physics (in Mach's correspondence around 1910, for example, with A. Föppl); according to F. Herneck[45], Ernst Mach specially invited the young Viennese physicist Philipp Frank to visit him "in order to find out more about the relativity theory, above all about the use of four-dimensional geometry". As a result, Frank, who had recently finished his studies under Ludwig Boltzmann and had begun to publish noteworthy contributions to relativity, published the "presentation of Einstein's theory to which Mach gave his assent" under the title 'Das Relativitätsprinzip und die Darstellung der physikalischen Erscheinungen im vierdimensionalen Raum'.[46] It is an attempt, addressed to readers "who do not master modern mathematical methods", to show that Minkowski's work brings out the "empirical facts far more clearly by the use of four-dimensional world lines". The essay ends with the reassuring conclusion: "In this four-dimensional world the facts of experience can be presented more adequately than in three-dimensional space, where always only an arbitrary and one-sided projection is pictured."

Following Minkowski's own papers on the whole, Frank's treatment can make it nevertheless still appear that in most respects the time dimension is equivalent to the space dimensions. Thereby one could think that Minkowski's treatment based itself not only on a functional and operational interconnection of space and time, but also – fully in accord with Mach's own views – on the primacy of ordinary, 'experienced' space and time in the relativistic description of phenomena.

Perhaps as a result of this presentation, Mach invoked the names of Lorentz, Einstein, and Minkowski in his reply of 1910 to Planck's first attack, citing them as physicists "who are moving closer to the problems of matter, space, and time". Already a year earlier, Mach seems to have been hospitable to Minkowski's presentation, although not without reservations. Mach wrote in the 1909 edition of *Conservation of Energy*[22]: "Space and time are here conceived not as independent entities, but as forms of the dependence of the phenomena on one another"; he also added a reference to Minkowski's lecture of 1908.[44] But a few lines earlier, Mach had written: "Spaces of many dimensions seem to me not so essential for physics. I would only uphold them if things of thought [*Gedankendinge*] like atoms are maintained to be indispensable, and if, then, also the freedom of working hypotheses is upheld."

It was correctly pointed out by C. B. Weinberg[47] that Mach may eventually have had two sources of suspicion against the Minkowskian form of relativity theory. As was noted above, Mach regarded the fundamental notions of mechanics as problems to be continually discussed with maximum openness within the frame of empiricism, rather than as questions that can be solved and settled – as the relativists, seemingly dogmatic and sure of themselves, were in his opinion more and more inclined to do. In addition, Mach held that the questions of physics were to be studied in a broader setting, encompassing biology and psychophysiology. Thus Mach wrote: "Physics is not the entire world; biology is there too, and belongs essentially to the world picture."[48]

But I see also a third reason for Mach's eventual antagonism against such conceptions as Minkowski's (unless one restricted their application to "mere things of thought like atoms and molecules, which by their very nature can never be made the objects of sensuous contemplations"[49]). If one takes Minkowski's essay seriously – for example, the abandonment of space and time separately, with identity granted only to 'a kind of union of the two' – one must recognize that it entails the abandonment of the conceptions of experiential space and experiential time; and that is an attack on the very roots of sensations-physics, on the meaning of actual measurements. If identity, meaning, or 'reality' lies in the four-dimensional space-time interval *ds*, one is dealing with a quantity which is hardly *denkökonomisch*, nor one that preserves the primacy of measurements in 'real' space and time. Mach may well have seen the warning flag; and worse was soon to come, as we shall see at once.

In his exuberant lecture of 1908 (see Ref. 44), Minkowski had announced that "three-dimensional geometry becomes a chapter in four-dimensional physics.... Space and time are to fade away into the shadows, and only *eine Welt an sich* will subsist". In this 'world' the crucial innovation is the conception of the '*zeitartige Vektorelement*', *ds*, defined as $(1/c)\sqrt{c^2 dt^2 - dx^2 - dy^2 - dz^2}$ with imaginary components. To Mach, the word *Element* had a crucial and very different meaning. As we saw in Schlick's summary, elements were nothing less than the sensations and complexes of sensations of which the world consists and which completely define the world. Minkowski's rendition of relativity theory was now revealing the need to move the ground of basic, elemental truths from the plane of direct experience in ordinary space and time to a mathematicized, formalistic model of the world in a union of space and time that is not directly accessible to sensation – and, in this respect, is reminiscent of absolute space and time concepts that Mach had called "metaphysical monsters".[50]

Here, then, is an issue which, from the beginning, had separated Einstein and Mach even before they realized it. To the latter, the fundamental task of science was economic and descriptive; to the former, it was speculative-constructive and intuitive. Mach had once written: "If all the individual facts – all the individual phenomena, knowledge of which we desire – were immediately accessible to us, science would never have arisen".[51] To this, with the forthrightness caused perhaps by his recent discovery of Mach's opposition, Einstein countered during his lecture in Paris of 6 April 1922: "Mach's system studies the existing relations between data of experience: for Mach, science is the totality of these relations. That point of view is wrong, and in fact what Mach has done is to make a catalog, not a system".[52]

We are witnessing here an old conflict, one that has continued throughout the development of the sciences. Mach's phenomenalism brandished an undeniable and irresistible weapon for the critical re-evaluation of classical physics, and in this it seems to hark back to an ancient position that looked upon sensuous appearances as the beginning and end of scientific achievement. One can read Galileo in this light, when he urges the primary need of *description* for the fall of bodies, leaving '*the causes*' to be found out later. So one can understand (or rather, misunderstand) Newton, with his too-well-remembered remark: "I feign no hypotheses".[53] Kirchhoff is in this tradition. Boltzmann wrote of him in 1888:

> The aim is not to produce bold hypotheses as to the essence of matter, or to explain the movement of a body from that of molecules, but to present equations which, free from hypotheses, are as far as possible true and quantitatively correct correspondents of the phenomenal world, careless of the essence of things and forces. In his book on mechanics, Kirchhoff will ban all metaphysical concepts, such as forces, the cause of a motion; he seeks only the equations which correspond so far as possible to observed motions.[54]

And so could, and did, Einstein himself understand the Machist component of his own early work.

Phenomenalistic positivism in science has always been victorious, but only up to a very definite limit. It is the necessary sword for destroying old error, but it makes an inadequate plowshare for cultivating a new harvest. I find it exceedingly significant that Einstein saw this during the transition phase of partial disengagement from the Machist philosophy. In the spring of 1917 Einstein wrote to Besso and mentioned a manuscript which Friedrich Adler had sent him. Einstein commented: "He rides Mach's poor horse to exhaustion". To this, Besso – the loyal Machist – responds on 5 May 1917: "As to Mach's little horse, we should not insult it; did it not make possible the infernal journey through the relativities? And who knows – in the case of the nasty quanta, it may also carry Don Quixote de la Einsta through it all!"

Einstein's answer of 13 May 1917 is revealing: "I do not inveigh against Mach's little horse; but you know what I think about it. It cannot give birth to anything living, it can only exterminate harmful vermin."

VI. TOWARD A RATIONALISTIC REALISM

The rest of the pilgrimage is easy to reconstruct, as Einstein more and more openly and consciously turned Mach's doctrine upside down – minimizing rather than maximizing the role of actual details of experience, both at the beginning and at the end of scientific theory, and opting for a rationalism that almost inevitably would lead him to the conception of an objective, 'real' world behind the phenomena to which our senses are exposed.

In the essay, 'Maxwell's Influence on the Evolution of the Idea of Physical Reality' (1931), Einstein began with a sentence that could have been taken almost verbatim from Max Planck's attack on Mach in 1909, cited above: "The belief in an external world independent of the perceiving subject is the basis of all natural science". Again and again, in the period beginning with his work on the general relativity theory, Einstein insisted that between experience and reason, as well as between the world of sensory perception and the objective world, there are logically unbridgeable chasms. He characterized the efficacy of reason to grasp reality by the word *miraculous*; the very terminology in these statements would have been anathema to Mach.

We may well ask when and under what circumstances Einstein himself became aware of his change. Here again, we may turn for illumination to one of the hitherto unpublished letters, one written to his old friend, C. Lanczos, on 24 January 1938:

> Coming from sceptical empiricism of somewhat the kind of Mach's, I was made, by the problem of gravitation, into a believing rationalist, that is, one who seeks the only trustworthy source of truth in mathematical simplicity. The logically simple does not, of course, have to be physically true; but the physically true is logically simple, that is, it has unity at the foundation.

Indeed, all evidence points to the conclusion that Einstein's work on general relativity theory was crucial in his epistemological development. As he wrote later in 'Physics and Reality' (1936): "the first aim of the general theory of relativity was the preliminary version which, while not meeting the requirements for constituting a closed system, could be connected in as simple a manner as possible with 'directly observed facts'". But the aim, still apparent during the first years of correspondence with Mach, could not be achieved. In notes on the origin of the general relativity theory,[55] Einstein reported:

> I soon saw that the inclusion of non-linear transformation, as the principle of equivalence demanded, was inevitably fatal to the simple physical interpretation of the coordinate – i.e., that it could no longer be required that coordinate differences [*ds*] should signify direct results of measurement with ideal scales or clocks. I was much bothered by this piece of knowledge ... [just as Mach must have been].
>
> The solution of the above mentioned dilemma [from 1912 on] was therefore as follows: A physical significance attaches not to the differentials of the coordinates, but only to the Riemannian metric corresponding to them.[55]

And this is precisely a chief result of the 1913 essay of Einstein and Grossmann[56], the same paper which Einstein sent to Mach and discussed in his fourth letter. This result was the final consequence of the Minkowskian four-space representation – the sacrifice of the primacy of direct sense perception in constructing a physically significant system. It was the choice that Einstein had to make – against fidelity to a catalogue of individual operational experiences and in favor of fidelity to the ancient hope for a unity at the base of physical theory.[57]

Enough has been written in other places to show the connections that existed between Einstein's scientific rationalism and his religious beliefs. Max Born summarized it in one sentence: "He believed in the power of reason to guess the laws according to which God has built the world".[58] Perhaps the best expression of this position by Einstein himself is to be found in his essay, 'Über den gegenwärtigen Stand der Feld-Theorie', in the *Festschrift* of 1929 for Aurel Stodola[59]:

> Physical Theory has two ardent desires, to gather up as far as possible all pertinent phenomena and their connections, and to help us not only to know *how* Nature is and *how* her transactions are carried through, but also to reach as far as possible the perhaps utopian and seemingly arrogant aim of knowing why Nature is *thus and not otherwise*. Here lies the highest satisfaction of a scientific person. ... [On making deductions from a "fundamental hypothesis" such as that of the kinetic-molecular theory], one experiences, so to speak, that God Himself could not have arranged those connections [between, for example, pressure, volume, and temperature] in any other way than that which factually exists, any more than it would be in His power to make the number 4 into a prime number. This is the promethean element of the scientific experience. ... Here has always been for me the particular magic of scientific considerations; that is, as it were, the religious basis of scientific effort.

This fervor is indeed far from the kind of analysis which Einstein had made only a few years earlier. It is doubly far from the asceticism of his first philosophic mentor, Mach, who had written in his day book: "Colors, space, tones, etc. These are the only realities. Others do not exist."[60] It is, on the contrary, far closer to the rational realism of his first scientific mentor Planck, who had written: "The disjointed data of experience can never furnish a veritable science without the intelligent interference of a spirit actuated by faith.... We have a right to feel secure in surrendering to our belief in a philosophy of the world based upon a faith in the rational ordering of this world".[61] Indeed, we note the philosophical kinship of Einstein's position with seventeenth-century natural philosophers – for example, with Johannes Kepler who, in the Preface of the *Mysterium Cosmographicum*, announced that he wanted to find out concerning the number, positions, and motions of the planets, "why they are as they are, and not otherwise", and who wrote to Herwart in April, 1599, that, with regard to numbers and quantity, "our knowledge is of the same kind as

God's, at least insofar as we can understand something of it in this mortal life".

Not unexpectedly, we find that during this period (around 1930) Einstein's non-scientific writings began to refer to religious questions much more frequently than before. There is a close relation between his epistemology, in which reality does not need to be validated by the individual's sensorium, and what he called "Cosmic religion"[62], defined as follows: "The individual feels the vanity of human desires and aims, and the nobility and marvelous order which are revealed in nature and in the world of thought. He feels the individual destiny as an imprisonment and seeks to experience the totality of existence as a unity full of significance."

Needless to say, Einstein's friends from earlier days sometimes had to be informed of his change of outlook in a blunt way. For example, Einstein wrote to Moritz Schlick on 28 November 1930:

> In general your presentation fails to correspond to my conceptual style insofar as I find your whole orientation so to speak too positivistic. ... I tell you straight out: Physics is the attempt at the conceptual construction of a model of the *real world* and of its lawful structure. To be sure, it [physics] must present exactly the empirical relations between those sense experiences to which we are open; but only *in this way* is it chained to them. ... In short, I suffer under the (unsharp) separation of Reality of Experience and Reality of Being. ...
>
> You will be astonished about the "metaphysicist" Einstein. But every four- and two-legged animal is de facto in this sense metaphysicist. [Emphasis in the original]

Similarly, P. Frank, Einstein's early associate and later his biographer, reports that the realization of Einstein's true state of thought reached Frank in a most embarrassing way, at the congress of German physicists in Prague in 1929, just as Frank was delivering "an address in which I attacked the metaphysical position of the German physicists and defended the positivistic ideas of Mach." The very next speaker disagreed and showed Frank that he had been mistaken still to associate Einstein's views with that of Mach and himself. "He added that Einstein was entirely in accord with Planck's view that physical laws describe a reality in space and time that is independent of ourselves. At that time", Frank comments, "this presentation of Einstein's views took me very much by surprise".[63]

In retrospect it is, of course, much easier to see the evidences that this change was being prepared. Einstein himself realized more and more clearly how closely he had moved to Planck, from whom he earlier dissociated himself in three of the four letters to Mach. At the celebration of Planck's sixtieth birthday, two years after Mach's death, Einstein made a moving speech in which, perhaps for the first time, he referred publicly to the Planck-Mach dispute and affirmed his belief "there is no logical way to the discovery of these elementary laws. There is only the way of intuition" based on *Einfühlung* in experience.[64] The scientific dispute concerning the theory of radiation between Einstein and Planck, too, had been settled (in Einstein's favor) by a sequence of developments after 1911 – for example, by Bohr's theory of radiation from gas atoms. As colleagues, Planck and Einstein saw each other regularly from 1913 on. Among evidences of the coincidence of these outlooks there is in the Einstein Archives a handwritten draft, written on or just before 17 April 1931 and intended as Einstein's introduction to Planck's hard-hitting article 'Positivism and the Real External World'.[65] In lauding Planck's article, Einstein concludes: "I presume I may add that both Planck's conception of the logical state of affairs as well as his subjective expectation concerning the later development of our science corresponds entirely with my own understanding".[66]

This essay gave a clear exposition of Planck's (and one may assume, Einstein's) views, both in physics and in philosophy more generally. Thus Planck wrote there:

> The essential point of the positivist theory is that there is no other source of knowledge except the straight and short way of perception through the senses. Positivism always holds strictly to that. Now, the two sentences: (1) *there is a real outer world which exists independently of our act of knowing* and (2) *the real outer world is not directly knowable* form together the cardinal hinge on which the whole structure of physical science turns. And yet there is a certain degree of contradiction between those two sentences. This fact discloses the presence of the irrational, or mystic, element which adheres to physical science as to every other branch of human knowledge. The effect of this is that a science is never in a position completely and exhaustively to solve the problem it has to face. We must accept that as a hard and fast, irrefutable fact, and this fact cannot be removed by a theory which restricts the scope of science at its very start. Therefore, we see the task of science arising before us as an incessant struggle toward a goal which will never be reached, because by its very nature it is unreachable. It is of a metaphysical character, and, as such, is always again and again beyond our achievement.[67]

From then on, Einstein's and Planck's writings on these matters are often almost indistinguishable from each other. Thus, in an essay in honor of Bertrand Russell[68], Einstein warns that the ' fateful fear of metaphysics'

... has come to be a malady of contemporary empiricistic philosophizing". On the other hand, in the numerous letters between the two old friends, Einstein and Besso, each to the very end touchingly and patiently tries to explain his position, and perhaps to change the other's. Thus, on 28 February 1952, Besso once more presents a way of making Mach's views again acceptable to Einstein. The latter, in answering on 20 March 1952, once more responds that the facts cannot lead to a deductive theory and, at most, can set the stage "for intuiting a general principle" as the basis of a deductive theory. A little later, Besso is gently scolded (in Einstein's letter of 13 July 1952): "It appears that you do not take the four-dimensionality of reality seriously, but that instead you take the present to be the only reality. What you call 'world' is in physical terminology 'spacelike sections' for which the relativity theory – already the special theory – denies objective reality."

In the end, Einstein came to embrace the view which many, and perhaps he himself, thought earlier he had eliminated from physics in his basic 1905 paper on relativity theory: that there exists an external, objective, physical reality which we may hope to grasp – not directly, empirically, or logically, or with fullest certainty, but at least by an intuitive leap, one that is only guided by experience of the totality of sensible 'facts'. Events take place in a 'real world', of which the space-time world of sensory experience, and even the world of multidimensional continua, are useful conceptions, but no more than that. For a scientist to change his philosophical beliefs so fundamentally is rare, but not unprecedented. Mach himself underwent a dramatic transformation quite early (from Kantian idealism, at about age seventeen or eighteen, according to Mach's autobiographical notes). We have noted that Ostwald changed twice, once to anti-atomism and then back to atomism. And strangely, Planck himself confessed in his 1910 attack on Mach (Ref. 23) that some twenty years earlier, near the beginning of his own career when Planck was in his late twenties (and Mach was in his late forties), he, too, had been counted "one of the decided followers of the Machist philosophy", as indeed is evident in Planck's early essay on the conservation of energy (1887).

In an unpublished fragment apparently intended as an additional critical reply to one of the essays in the collection *Albert Einstein: Philosopher-Scientist* (1949), Einstein returned once more – and quite scathingly – to deal with the opposition. The very words he used showed how complete was the change in his epistemology. Perhaps even without consciously remembering Planck's words in the attack on Mach of 1909 cited earlier – that a basic aim of science is "the complete liberation of the physical world picture from the individuality of the separate intellects" [23] – Einstein refers to a 'basic axiom' in his own thinking:

It is the postulation of a "real world" which so-to-speak liberates the "world" from the thinking and experiencing subject. The extreme positivists think that they can do without it; this seems to me to be an illusion, if they are not willing to renounce thought itself.

Einstein's final epistemological message was that the world of mere experience must be subjugated by and based in fundamental thought so general that it may be called cosmological in character. To be sure, modern philosophy did not gain thereby a major, novel, and finished corpus. Physicists the world over generally feel that today one must steer more or less a middle course in the area between, on the one hand, the Machist attachment to empirical data or heuristic proposals as the sole source of theory and, on the other, the aesthetic-mathematical attachment to persuasive internal harmony as the warrant of truth. Moreover, the old dichotomy between rationalism and empiricism is slowly being dissolved in new approaches[69].

Yet by encompassing in his own philosophical development both ends of this range, and by always stating forthrightly and with eloquence his redefined position, Einstein had helped us to define our own.

Jefferson Laboratory,
Harvard University

REFERENCES

* Professor's Holton's essay first appeared in *Daedalus* **97** (Spring 1968) pp. 636–673, and is reprinted here by permission of the American Academy of Arts and Sciences.

1 Another recently published chapter for this study is the analysis of the probable sources of Einstein's first paper on relativity theory; see *Am. Scholar* **37** (Winter, 1967), 59–79. See also 'Einstein, Michelson, and the 'Crucial' Experiment', *Isis* **60** (Summer 1969), 132–197.

2 These documents are mostly on deposit at the Archives of the Estate of Albert Einstein at Princeton; where not otherwise indicated, citations made here are from those documents. In studying and helping to order for scholarly purposes the materials in the Archives, I have benefited from and am grateful for the help received from the Trustees of the Albert Einstein Estate, and particularly from Miss Helen Dukas. – I thank the Executor of the Estate for permission to quote from the writings of Albert Einstein. I also wish to acknowledge the financial support provided by the

Rockefeller Foundation for cataloguing the collection in the Archives at Princeton. The Institute for Advanced Study at Princeton and its director have been most hospitable throughout this continuing work. I am also grateful to M. Vero Besso for permission to quote from the letters of his father, Michelange Besso. All translations here are the author's, unless otherwise indicated. – Early drafts of portions of this essay have been presented as invited papers at the *Tagung* of *Eranos* in Ascona (August, 1965), at the International Congress for the History of Science in Warsaw (August, 1965), and at the meeting, *Science et Synthèse*, at UNESCO in Paris (December, 1965).

[3] F. Herneck, *Forschungen und Fortschritte* **36** (1964), 75. This and the next letter cited have been published by F. Herneck.

[4] The only other known attempt on Einstein's part to obtain an assistantship at that time was a request to Kammerlingh-Onnes (12 April 1901), to which, incidentally, he also seems to have received no response.

[5] J. T. Merz, *A History of European Thought in the Nineteenth Century*, Vol. II (reprint, Dover Publishing Co., N. Y., pp. 184, 199.

[6] 'Erinnerungen an Albert Einstein', issued by the Patent Office in Berne, about 1965 (n.d., no pagination).

[7] Wilhelm Ostwald, *Allgemeine Chemie*, **2** (2d ed.; 1893), p. 1014.

[8] *Albert Einstein: Philosopher-Scientist* (ed. by P. A. Schilpp), Evanston, Ill., 1946.

[9] *Die Mechanik in ihrer Entwicklung, historisch-kritisch dargestellt*, Leipzig 1883.

[10] In a special supplement on Ernst Mach in the journal *Neue Freie Presse* (Vienna), 12 June 1926.

[11] Einstein himself, in a brief and telling analysis, published also in the *Neue Freie Presse* of Vienna on 12 June 1926 (the day of unveiling of a monument to Mach), wrote: "Ernst Mach's strongest driving force was a philosophical one: the dignity of all scientific concepts and statements rests solely in isolated experiences [*Einzelerlebnisse*] to which the concepts refer. This fundamental proposition exerted mastery over him in all his research, and gave him the strength to examine the traditional fundamental concepts of physics (time, space, inertia) with an independence which at that time was unheard of".

[12] Among many evidences of Mach's effectiveness, not the least are his five hundred or more publications (counting all editions – for example, seven editions of his *The Science of Mechanics* in German alone during his lifetime), as well as his large exchange of letters, books, and reprints (of which many important ones "carry the dedication of their authors", to cite the impressive catalogue of Mach's library by Theodor Ackermann, Munich, No. 634 [1959] and No. 636 [1960]). A glimpse of Mach's effect on those near him was furnished by William James, who in 1882 heard Mach give a 'beautiful' lecture in Prague. Mach received James "with open arms. ... Mach came to my hotel and I spent four hours walking and supping with him at his club, an unforgettable conversation. I don't think anyone ever gave me so strong an impression of pure intellectual genius. He apparently has read everything and thought about everything, and has an absolute simplicity of manner and winningness of smile when his face lights up, that are charming". From James's letter, in Gay Wilson Allen, *William James, a Biography*, New York 1967, p. 249.

[13] Albert Einstein, 'Ernst Mach', *Phys. Z.* **17** (1916), 2.

[14] Gerald Holton, 'On the Origins of the Special Theory of Relativity', *Am. J. Phys.* **28** (October, 1960), 627–36.

[15] Gerald Holton, 'Note on the Thematic Analysis of Science: The Case of Poincaré and Relativity', *Mélanges Koyré* (Paris, 1964), pp. 257–268.

[16] For evidences that this insistence on prior epistemological analysis of conceptions of space and time are Machist rather than primarily derived from Hume and Kant (who had, however, also been influential), see Einstein's detailed rendition of Mach's critique of Newtonian space and time, in Ref. 13; his discussion of Mach in the *Autobiographical Notes*, pp. 27–29; and in Ref. 1.

[17] Albert Einstein, 'Zur Elektrodynamik bewegter Körper', *Ann. Physik*, **17** (1905), 893.

[18] See also P. Frank, in *Albert Einstein: Philosopher-Scientist* (ed. by P. A. Schilpp), pp. 272–73: "The definition of simultaneity in the special theory of relativity is based on Mach's requirement that every statement in physics has to state relations between observable quantities. ... There is no doubt that ... Mach's requirement, the 'positivistic' requirement, was of great heuristic value to Einstein."

[19] For example, see Philipp Frank, *Modern Science and Its Philosophy*, New York 1955, pp. 61–89; V. Kraft, *The Vienna Circle*, New York 1953; R. von Mises, *Ernst Mach und die Empiristische Wissenschaftsauffassung* (1938; printed as a fascicule of the series *Einheitswissenschaft*). [Eng. transl. in this volume – Ed.]

[20] J. Petzoldt, 'Gesellschaft für Positivistische Philosophie', reprinted in *Zeitschrift für Positivistische Philosophie*, **1** (1913), 4. In that same speech, Petzoldt sounded a theme that became widely favored in the positivistic interpretation of the genesis of relativity theory – namely, that the relativity theory was developed in direct response to the puzzle posed by the results of the Michelson experiment: "Clarity of thinking is inseparable from knowledge of a sufficient number of individual cases for each of the concepts used in investigation. Therefore, the chief requirement of positivistic philosophy: greatest respect for the facts. The newest phase of theoretical physics gives us an exemplary case. There, one does not hesitate, for the sake of a single experiment, to undertake a complete reconstruction. The Michelson experiment is the cause and chief support of this reconstruction, namely, the electrodynamic theory of relativity. To do justice to this experiment one has no scruples to submit the foundations of theoretical physics as it has hitherto existed, namely, Newtonian mechanics, to a profound transformation." – In his interesting essay 'Das Verhältnis der Machschen Gedankenwelt zur Relativitätstheorie', published as an appendix in the eighth German edition of Mach's *The Science of Mechanics* in the year 1921, Petzoldt faithfully attempts to identify and discuss several Machist aspects of Einstein's relativity theory: – (1) The theory "in the end is based on the recognition of the coincidence of sensations; and therefore it is fully in accord with Mach's worldview, which may be best characterized as a relativistic positivism" (p. 516). – (2) Mach's works "produced the atmosphere without which Einstein's Relativity Theory would not have been possible" (p. 494), and in particular Mach's analysis of the equivalence of rotating reference objects in Newton's bucket experiment prepared for the next step, Einstein's "equivalence of relatively moving coordinate systems" (p. 495). – (3) Mach's principle of economy is said to be marvellously exhibited in Einstein's succinct and simple statements of the two fundamental hypotheses. The postulate of the equivalence of inertial coordinate systems deals with "the simplest case thinkable, which now also serves as a fundamental pillar for the General Theory. And Einstein chose also with relatively greatest simplicity the other basic postulate [constancy of light velocity]. ... These are the foundations. Everything else is logical consequence" (pp. 497–98).

[21] Letter of Besso to Einstein, 16 February 1939. Among many testimonies to the effect of Einstein on positivistic philosophies of science, see P. W. Bridgman, 'Einstein's Theory and the Operational Point of View', *Albert Einstein: Philosopher-Scientist*, ed. by P. A. Schilpp.

[22] Ernst Mach, *History and Root of the Principle of the Conservation of Energy* (Chicago, 1911), translation by P. Jourdain of Second Edition (1909), p. 95. For a brief analysis of Mach's various expressions of adherence as well as reservations with respect to the principle of relativity, see H. Dingler, *Die Grundlagen der Machschen Philosophie* (Leipzig, 1924), pp. 73–86. – F. Herneck (*Phys. Bl.* **17** (1961), 276) reports that P. Frank wrote him he had the impression during a discussion with Ernst Mach around 1910 that Mach "was fully in accord with Einstein's special relativity theory, and particularly with its philosophical basis".

[23] Republished in M. Planck, *A Survey of Physical Theory* (New York, 1960), p. 24. We shall read later a reaffirmation of this position, in almost exactly the same words, but from another pen. – After Mach's rejoinder (*Scientia* **7** (1910), 225), Planck wrote a second, much more angry essay, 'Zur Machschen Theorie der physikalischen Erkenntnis', *Vierteljahrschrift für wissenschaftliche Philosophie* **34** (1910), 497. He ends as follows: "If the physicist wishes to further his science, he must be a Realist, not an Economist [in the sense of Mach's principle of economy]; that is, in the flux of appearances he must above all search for and unveil that which persists, is not transient, and is independent of human senses."

[24] Later Einstein found that this procedure did not work; see *Ideas and Opinions* (New York, 1954), p. 286, and other publications. In a letter of 2 February 1954 to Felix Pirani, Einstein writes: "One shouldn't talk at all any longer of Mach's principle, in my opinion. It arose at a time when one thought that 'ponderable bodies' were the only physical reality and that in a theory all elements that are fully determined by them should be conscientiously avoided. I am quite aware of the fact that for a long time, I, too, was influenced by this fixed idea."

[25] *Z. Math. Phys.* **62** (1913), 225–61.

[26] For a further analysis and the full text of the four letters, see F. Herneck, *Forschungen und Fortschritte*, **37** (1963), 239–43, and *Wissenschaftliche Zeitschrift der Friedrich-Schiller-Universität Jena*, **15** (1966), 1–14; and H. Hönl, *Phys. Bl.* **16** (1960), 571. Many other evidences, direct and indirect, have been published to show Mach's influence on Einstein prior to Mach's death in 1916. For example, recently a document has been found which shows that in 1911 Mach had participated in formulating and signing a manifesto calling for the founding of a society for the positivistic philosophy. Among the signers, together with Mach, we find Joseph Petzoldt, David Hilbert, Felix Klein, Georg Helm, Sigmund Freud, and Einstein. (See F. Herneck, *Phys. Bl.* **17** (1961), 276.)

[27] *Phys. Z.* **11** (1910), 605. To be sure, Mach was not at all times a rigid "Machist" himself.

[28] See Albert Einstein, 'Time, Space, and Gravitation' (1948), *Out of My Later Years* (New York, 1950). Einstein makes the distinction between constructive theories and "theories of principle". Einstein cites, as an example of the latter, the relativity theory, and the laws of thermodynamics. Such theories of principle, Einstein says, start with "empirically observed general properties of phenomena". See also *Autobiographical Notes*, p. 53, and *The World as I see it*, pp. 74–75.

[29] *Bull. Soc. Franc. Phil.* **22** (Paris, 1922), 111. In his 1913 preface rejecting relativity, Mach expressed himself perhaps more impetuously and irascibly than he may have meant. Some evidence for this possibility is in Mach's letters to J. Petzoldt. On 27 April 1914 Mach wrote: "I have received the copy of the positivistic *Zeitschrift* which contains your article on relativity; I liked it not only because you copiously acknowledge my humble contributions with respect to that theme, but also in general." And on 1 May 1914 Mach writes – rather more incoherently – to Petzoldt: "The enclosed letter

Einstein [a copy of the last of Einstein's four letters, cited above] proves the penetration of positivistic philosophy into physics; you can be glad about it. A year ago, philosophy was altogether sheer nonsense. The details prove it. The paradox of the clock would not have been noticed by Einstein a year ago."

I thank Mr. John Blackmore for drawing my attention to the Mach–Petzoldt letters, and to Dr. H. Müller for providing copies from the Petzoldt Archive at the Technische Hochschule in Berlin.

[30] R. S. Shankland, *Am. J. Phys.* **31** (1963), 56.

[31] A typical example is a letter of 18 September 1930 to Armin Weiner: "... I did not have a particularly important exchange of letters with Mach. However, Mach did have a considerable influence upon my development through his writings. Whether or to what extent my life's work was influenced thereby is impossible for me to find out. Mach occupied himself in his last years with the relativity theory, and in a preface to a late edition of one of his works even spoke out in rather sharp refusal against the relativity theory. However, there can be no doubt that this was a consequence of a lessening ability to take up [new ideas] owing to his age, for the whole direction of thought of this theory conforms with Mach's, so that Mach quite rightly is considered as a forerunner of general relativity theory. ..." – I thank Colonel Bern Dibner for making a copy of the letter available to me from the Archives of the Burndy Library in Norwalk, Connecticut. Among other hitherto unpublished letters in which Einstein indicated his indebtedness to Mach, we may cite one to A. Lampa, 9 December 1935: "... You speak about Mach as about a man who has gone into oblivion. I cannot believe that this corresponds to the facts since the philosophical orientation of the physicists today is rather close to that of Mach, a circumstance which rests not a little on the influence of Mach's writings." – Moreover, practically everyone else shared Einstein's explicitly expressed opinion of the debt of relatively theory to Mach; thus H. Reichenbach wrote in 1921 (*Logos* **10**, 331): "Einstein's theory signifies the accomplishment of Mach's program." Even Hugo Dingler agreed: "[Mach's] criticism of the Newtonian conceptions of time and space served as a starting point for the relativity theory. ... Not only Einstein's work, but even more recent developments, such as Heisenberg's quantum mechanics, have been inspired by the Machian philosophy." *Encyclopedia of the Social Sciences*, vol. IX (New York, 1933), p. 653. And H. E. Hering wrote an essay whose title is typical of many others: 'Mach als Vorläufer des physikalischen Relativitätsprinzips', *Kölner Universitätszeitung* **1** (January 17, 1920), 3–4. I thank Dr. J. Blackmore for a copy of the article.

[32] In the special supplement of *Neue Freie Presse* of Vienna on 12 June 1926, Einstein – then already disenchanted for some time with the Machist program – wrote immediately after the portion quoted in Reference 11: "Philosophers and scientists have often criticized Mach, and correctly so, because he erased the logical independence of the concepts vis-à-vis the 'sensations', [and] because he wanted to dissolve the Reality of Being, without whose postulation no physics is possible, in the Reality of Experience. ..." – There are additional resources, both published and unpublished, on the detailed aspects of the relation between Einstein and Mach, which, for lack of space, cannot be summarized here.

[33] For example, by Einstein himself, by Joseph Petzoldt, and by Hugo Dingler (cf. H. Dingler, *Encyclopedia of the Social Sciences*, pp. 84–86). I assign relatively little weight to the possibility that the rift grew out of the difference between Einstein and Mach on atomism. Herneck provides the significant report that according to a letter from P. Frank, Mach was personally influenced by Dingler, whom Mach had praised

in the 1912 edition of the *Mechanik* and who was from the beginning an opponent of relativity theory, becoming one of the most "embittered enemies" of Einstein (F. Herneck, *op. cit.* [Ref. 26], p. 14). The copies of letters from Dingler to Mach in the Ernst-Mach-Institut in Freiburg indicate Dingler's intentions; nevertheless, there remains a puzzle about Dingler's role which is worth investigating. It is significant that in his 1921 essay, Petzoldt (Ref. 20) devotes much space to a defense of Einstein's work against Dingler's attacks. See also Joachim Thiele's detailed analysis of Mach's Preface, in *NTM, Schriftenreihe für Geschichte der Naturwissenschaften, Technik und Medizin*, vol. II (Leipzig, 1965), pp. 10–19.

[34] 'On the Theory of Relativity', *Mein Weltbild* (Amsterdam, 1934); republished in *Ideas and Opinions* (New York, 1954), p. 246. F. Herneck has given the texts of similar discussions on phonographic records by Einstein in 1921 and even in 1924; cf. *Forschungen und Fortschritte*, **40** (1966), pp. 133–34.

[35] Quotations from 'On the Method of Theoretical Physics', *Mein Weltbild* (1934), as reprinted in translation in *Ideas and Opinions*, pp. 270–76, except for correction of mistranslation of one line. There are a number of later lectures and essays in which the same point is made. See, for example, the lecture, 'Physics and Reality' (1936, reprinted in *Ideas and Opinions*), which states that Mach's theory of knowledge is insufficient on account of the relative closeness between experience and the concepts which it uses; Einstein advocates going beyond this 'phenomenological physics' to achieve a theory whose basis may be further removed from direct experience, but which in return has more 'unity in the foundations'. Or see *Autobiographical Notes*, p. 27: "In the choice of theories in the future", he indicates that the basic concepts and axioms will continue to "distance themselves from what is directly observable". – Even as Einstein's views developed to encompass the '*erlebbare, beobachtbare*' facts as well as the '*wild-spekulative*' nature of theory, so did those of many of the philosophers of science who also had earlier started from a more strict Machist position. This growing modification of the original position, partly owing to "the growing understanding of the general theory of relativity", has been chronicled by P. Frank, for example, in 'Einstein, Mach, and Logical Positivism', in *Albert Einstein: Philosopher-Scientist* (ed. by P. A. Schilpp).

[36] For a discussion of thematic and phenomenic elements in theory construction, see G. Holton, 'The Thematic Imagination in Science', in *Science and Culture* (ed. by G. Holton), Boston, 1965.

[37] P. A. Schilpp (ed.), *Albert Einstein: Philosopher-Scientist*, p. 53. Emphases added. On pp. 9–11, Einstein describes what may be a possible precursor of this attitude in his study of geometry as a child.

[38] W. Kaufmann, 'Concerning the Constitution of the Electron', *Ann. Physik* **19** (1906), 495. Emphasis in original.

[39] *Jahrbuch der Radioaktivität und Elektronik*, **4** (1907), 28. Shortly after Kaufmann's article appeared, M. Planck (in *Phys. Z.* **7** (1906), 753–61) took it on himself publicly to defend Einstein's work in an analysis of Kaufmann's claim. He concluded that Kaufmann's data did not have sufficient precision for his claim. Incidentally, Planck tried to coin the term for the new theory that had not yet been named: 'Relativtheorie'.

[40] It should be remembered that Poincaré, with a much longer investment in attempts to fashion a theory of relativity, was quite ready to give in to the experimental 'evidence'. See Ref. 15.

[41] Carl Seelig, *Albert Einstein* (Zürich, 1960), p. 195.

[42] Albert Einstein, 'Notes on the Origin of the General Theory of Relativity', *Mein*

Weltbild (Amsterdam, 1934), reprinted in translation in *Ideas and Opinions* (New York, 1954), pp. 285–90.

43 "*Da könnt' mir halt der liebe Gott leid tun, die Theorie stimmt doch.*" This semi-serious remark of a person who was anything but sacriligious indeed illuminates the whole style of a significant group of new physicists. P. A. M. Dirac, in *Scientific Am.* **208** (May 1963), 47–48, speaks about this, with special attention to the work of Schrödinger, a spirit close to that of his friend, Einstein, despite the ambivalence of the latter to the advances in quantum physics. We can do no better than quote *in extenso* from Dirac's account: "Schrödinger worked from a more mathematical point of view, trying to find a beautiful theory for describing atomic events, and was helped by deBroglie's ideas of waves associated with particles. He was able to extend deBroglie's ideas and to get a very beautiful equation, known as Schrödinger's wave equation, for describing atomic processes. Schrödinger got this equation by pure thought, looking for some beautiful generalization of deBroglie's ideas, and not by keeping close to the experimental development of the subject in the way Heisenberg did. – I might tell you the story I heard from Schrödinger of how, when he first got the idea for this equation, he immediately applied it to the behavior of the electron in the hydrogen atom, and then he got results that did not agree with experiment. The disagreement arose because at that time it was not known that the electron has a spin. That, of course, was a great disappointment to Schrödinger, and it caused him to abandon the work for some months. Then he noticed that if he applied the theory in a more approximate way, not taking into account the refinements required by relativity, to this rough approximation his work was in agreement with observation. He published his first paper with only this rough approximation, and in this way Schrödinger's wave equation was presented to the world. Afterward, of course, when people found out how to take into account correctly the spin of the electron the discrepancy between the results of applying Schrödinger's relativistic equation and the experiments was completely cleared up. – I think there is a moral to this story, namely, that it is more important to have beauty in one's equations than to have them fit experiment. If Schrödinger had been more confident of his work, he could have published it some months earlier, and he could have published a more accurate equation. That equation is now known as the Klein-Gordon equation, although it was really discovered by Schrödinger, and in fact was discovered by Schrödinger before he discovered his non-relativistic treatment of the hydrogen atom. It seems that if one is working from the point of view of getting beauty in one's equations, and if one has really a sound insight, one is on a sure line of progress. If there is not complete agreement between the results of one's work and experiment, one should not allow oneself to be too discouraged, because the discrepancy may well be due to minor features that are not properly taken into account and that will get cleared up with further developments of the theory. That is how quantum mechanics was discovered ..." [pp. 46–47].

44 Published several times – for example, by Teubner, Leipzig, 1909.

45 F. Herneck, *Phys. Bl.* **15** (1959), 565. P. Frank's remark is reported by F. Herneck, in *Ernst Mach* (ed. by W. F. Merzkirch), Freiburg, 1967, p. 50.

46 *Z. Phys. Chem.* (1910), pp. 466–95.

47 Mach's Empirio-Pragmatism in Physical Science, Thesis, Columbia University, 1937.

48 Ernst Mach, *Scientia* **7** (1910), 225.

49 *Space and Geometry* (1906), p. 138. Mach's attempts to speculate on the use of *n*-dimensional spaces for representing the configuration of such "mere things of

thought" – the derogatory phrase also applied to absolute space and absolute motion in Newton – are found in his first major book, *Conservation of Energy* (first edition, 1872).

[50] Cf. J. Petzoldt, 'Verbietet die Relativitätstheorie Raum und Zeit als etwas Wirkliches zu denken?', *Verhandlungen der Deutschen Physikalischen Gesellschaft*, Nos. 21–24 (1918), pp. 189–201. Here, and again in his 1921 essay (Ref. 20), Petzoldt tries to protect Einstein from the charge – for example, that by Sommerfeld – that space and time no longer "are to be thought of as real".

[51] Ernst Mach, *Conservation of Energy*, p. 54.

[52] *Op. cit.* Ref. 29, also reported in 'Einstein and the Philosophies of Kant and Mach', *Nature*, **112** (August 1923), 253.

[53] That Einstein did not so misunderstand Newton can be illustrated, for example, in a comment reported by C. B. Weinberg: "Dr. Einstein further maintained that Mach, as well as Newton, tacitly employs hypotheses – not recognizing their non-empirical foundations." (Weinberg, *op. cit.*, p. 55.) For an example of Mach's tacit presuppositions, see H. Dingler, *Encyclopedia of the Social Sciences*, pp. 69–71. Dingler also analyzed some of the non-empirical foundations of relativity theory in *Kritische Bemerkungen zu den Grundlagen der Relativitätstheorie*, Leipzig 1921.

[54] Cited by R. S. Cohen in his very useful essay in P. Schilpp (ed.), *The Philosophy of Rudolf Carnap*, La Salle, Ill., 1963, p. 109. I am also grateful to Professor Cohen for a critique of parts of this paper in earlier form.

[55] Albert Einstein, 'Notes on the Origin of the General Relativity Theory', pp. 288–89.

[56] Ref. 25, pp. 230–31.

[57] I am not touching in this essay on the effect of quantum mechanics on Einstein's epistemological development; the chief reason is that while from his 'heuristic' announcement of the value of a quantum theory in 1905 Einstein remained consistently skeptical about the 'reality' of the quantum theory of radiation, this opinion only added to the growing realism stemming from his work on general relativity theory. In the end, he reached the same position in quantum physics as in relativity; cf. his letter of 7 September 1944 to Max Born: "In our scientific expectations we have become antipodes. You believe in the dice-playing God, and I in perfect rules of law in a world of something objectively existing, which I try to catch in a wildly speculative way." (Reported by Max Born, *Universitas* **8** (1965), 33.)

[58] Max Born, 'Physics and Relativity', *Physics in My Generation*, London, 1956, p. 205.

[59] Orell Füssli Verlag, Zürich and Leipzig, 1929, pp. 126–32. I am grateful to Professor C. Lanczos and Professor John Wheeler for pointing out this reference to me.

[60] H. Dingler, *Die Grundlagen der Machschen Philosophie*, p. 98.

[61] *The Philosophy of Physics*, New York, 1936, pp. 122, 125.

[62] In 'Religion and Science', written for *The New York Times Magazine* 9 November 1930; cf. *Mein Weltbild*, p. 39, and *Cosmic Religion*, New York, 1931, p. 48. Possible reasons for Einstein's growing interest in these matters, partly related to the worsening political situation at the time, are discussed in P. Frank, *Einstein, His Life and Times*. It is noteworthy that while Einstein was quite unconcerned with religious matters during the period of his early scientific publications, he gradually returned later to a position closer to that at a very early age, when he reported he had felt a "deep religiosity. ... It is quite clear to me that the religious paradise of youth ... was a first attempt to free myself from the chains of the 'merely personal'". *Autobiographical Notes*, p. 5. For a discussion, see G. Holton, *The Graduate Journal*, **7** (Spring 1967) [University of Texas], 417–20.

[63] P. Frank, *Einstein, His Life and Times*, New York 1947, p. 215. Einstein's change of mind was, of course, not acceptable to a considerable circle of previously sympathetic scientists and philosophers. See for example P. W. Bridgman, 'Einstein's Theory and the Operational Point of View', in *Albert Einstein: Philosopher-Scientist* (ed. by P. A. Schilpp).
[64] Originally entitled 'Motiv des Forschens' in *Zu Max Planck*, Planck's 60. Geburtstag, Müller, Karlsruhe, 1918, the talk had a fate not untypical of many of Einstein's essays. It was reprinted without date or source, under the title 'Prinzipien der Forschung', in *Mein Weltbild*, and again, in perhaps unauthorized extension and translation by James Murphy, as a preface to M. Planck, *Where Is Science Going*? (London, 1933). In an earlier appreciation of Planck in 1913, Einstein had written only very briefly about Planck's epistemology, merely lauding Planck's essay of 1896 against energetics, and not mentioning Mach.
[65] *The International Forum*, vol. I (1931), Nos. 1–2.
[66] Einstein sent his introduction to the editor of the journal on 17 April 1931, but it appears to have come too late for inclusion.
[67] Max Planck, 'Positivism and External Reality', *The International Forum*, vol. I (Berlin 1931), Nos. 1–2, pp. 15–16. Emphasis in original.
[68] In *The Philosophy of Bertrand Russell* (ed. by P. Schilpp), Evanston 1944, p. 289.
[69] Toward the end, Einstein himself acknowledged a similar point in his 'Reply to Criticisms', in P. Schilpp (ed.), *op. cit.*, pp. 679–680: "'Einstein's position ... contains features of rationalism and extreme empiricism ....' This remark is entirely correct .... A wavering between these extremes appears to me unavoidable."

HUBERT GOENNER

MACH'S PRINCIPLE AND EINSTEIN'S THEORY OF GRAVITATION

I. INTRODUCTION

The increased interest of physicists in Einstein's theory of gravitation over the past years has been accompanied by a number of discussions of Mach's suggestion concerning a relation between inertial forces and distant stars. There has been, however, no accord on the mathematical formulation of Mach's ideas and on their meaning in General Relativity. My purpose is to review, in the framework of the present-day understanding of Einstein's theory, the changes in interpretation which Mach's ideas have experienced and to sort out different statements collectively labeled as Mach's principle.

II. MACH'S IDEAS

Mach criticized Newton's concept[1] of motion with respect to absolute space for being unobservable.[2] He substituted for it the concept of relative motion, that is motion with respect to the distant stars representing, at his time, the bulk of matter in the observable universe[3]. This amounts to pointing out a material system (distant stars) as an approximate realization of a mathematical construct (absolute space)[4]. Mach went on to assert, if only provisionally, that the motion of every single body is determined by all the other bodies in the universe.[5] The crucial point here is a change in the role of the distant stars: from being marker-points for a frame of reference to becoming part of an interacting physical system. Consequently, the agreement of the dynamical and astronomical measurements of the rotation of the earth is no longer trivial. This reevaluation of the status of distant stars by Mach has been criticized as mixing of the ontological and epistemological concepts of determination.[6] On the contrary, it has been taken as fundamental by others, including Einstein himself. Nowadays, Mach's ideas are understood to mean that the local inertial frame is defined[7,8] or determined[9,10] by or, more vaguely, just depends[10a] on distant matter (distant astronomical objects, the Galaxy,

Boston Studies in the Philosophy of Science, VI.

extragalactic nebulae, distribution of matter in the universe, etc.) A more specific formulation expresses the observation that the local inertial system is approximately irrotational with respect to the distant stars.[11] An equivalent statement, though given in a rather technical language, is that the local inertial compass and the light compass (of the fixed stars) must coincide. In addition, generalizations are to be found calling 'Machian' each connection between the local dynamics and the structure of the universe as a whole[12], as well as weaker statements linking the inertial behavior of a body only with the matter-distribution in its neighborhood.[13] Finally, concepts such as inertial force, inertial mass and inertia of a body are also used in statements about Mach's ideas.

III. INERTIAL SYSTEMS IN EINSTEIN'S THEORY

Inertial systems may be described either as mathematical constructs (mathematical reference systems) or as physical constructs (physical reference systems). The former consist of a whole *class* of reference systems, whereas the latter represent a single reference frame. For example, physicists are realizing a physical inertial system by the axes of a gyroscope and astronomers by listing the relative positions of neighboring stars. At first glance, the roles of inertial systems in Newton's and Einstein's theories of gravitation appear to be quite different. However, a careful analysis has shown that both theories may be described by a 4-dimensional space-time manifold the essential distinction being that Einstein's theory is metric whereas Newton's is only affine.[14,15] In such a formulation, in both theories the mathematical construction of the class of inertial systems used before now applies only *locally*. The local inertial systems are selected by the choice of suitable global coordinates in the *tangent* space. In the covariant formulation of Newton's theory one may nonetheless always find global coordinate systems in which the 3-geometry is Euclidian whereas Einstein's theory does not admit global coordinates reducing the metric to the Minkowskian values.

It is obvious that the mathematical construction of the *local* inertial systems is independent of any relation between the 4-geometry of space-time and the distribution of matter in it.

One might be tempted to define *global* inertial systems in Einstein's theory by help of isometric mappings on the Riemannian manifold which would allow the selection of privileged global coordinate systems. In the precious few cases, however, in which such a group of isometries exists, it will be different from the Lorentz group connecting the local inertial systems. Locally and globally selected reference systems will not coincide (for example in a space which admits the De Sitter group).

Another way of defining privileged coordinate systems would be to require a group of isometries only asymptotically, e.g. at spatial infinity (Section VIIIe). If one wants to arrive at the Lorentz group connecting the local inertial systems this implies then that the manifold is asymptotically flat. Apart from the fact that such an assumption is rather restrictive, the inertial system again would not be defined relative to distant masses.

Next, we consider the *operational* description of local inertial systems in Einstein's theory. As has been shown in detail[11,16,17] a local inertial system can be constructed by local dynamical experiments. For example, an observer moving along a time-like world line throws out at some instant a cloud of then freely falling test particles. The motion of such test particles is irrotational if referred to a tetrad obtained by Fermi-propagation along the observer's world line of his 3-dimensional rest frame. This tetrad represents the inertial compass. In this discussion the word 'local' cannot be taken in its strict mathematical sense, namely as referring to a single point. It describes a neighborhood of the observer, if only an arbitrarily small one.

For the operational definition of the light compass one observes the intersection of null rays coming from distant objects with a unit sphere surrounding the observer. If the points of intersection stay fixed during the motion of the observer (with regard to the inertial compass) the light compass and inertial compass are said to coincide.

It is clearly seen that this cannot hold in all circumstances in which a Fokker precession would arise.[18] (For example, it does not if the observer moves along a geodesic around a condensation of matter in a Friedmann model.) It also cannot be valid if space-time is anisotropic. Therefore, one would consider the coincidence of inertial and light compass to be the exception rather than the rule in Einstein's theory.

In addition to this, there are difficulties with the definition of the light compass of distant matter for solutions of Einstein's equations with an event horizon.

IV. EINSTEIN'S IDEAS

One has to discern two different lines of thought in Einstein's writings about Mach's ideas.[19] The first refers directly to the criticism of Newton's absolute space: "... a material particle does not move in unaccelerated motion relatively to space, but relatively to the centre of all the other masses in the universe. ..."[20] (cosmological aspect). On the other hand, one finds a purely local interpretation of Mach's ideas in Einstein's hypothesis of the relativity of inertia, viz. "that inertia depends on a mutual action of bodies".[21] Having substituted the gravitational interaction for this mutual action, he expected three effects to show up: (1) The inertia of a body must increase when ponderable masses are piled up in its neighborhood; (2) A body must experience an accelerating force when neighboring masses are accelerated; and (3) A rotating hollow body must generate inside of itself a Coriolis field which deflects moving bodies in the sense of rotation and a centrifugal field as well.[22] All three effects appear in Einstein's discussion of the equation of motion of a test-particle.[23] Since then, they have been used as criteria for Machian behavior of solutions of Einstein's equations, the third one being stressed especially: the dragging along of inertial systems by rotating masses.

V. RELATIVITY OF ROTATION

Newton exemplified his concept of motion with respect to absolute space by the well-known experiment with a rotating pail of water.[24] Ever since, in the criticism of Newton and in the formulation of Mach's ideas, special emphasis has been given to the so-called principle of the relativity of rotation, which states that there can be no absolute rotation, but only rotation with regard to matter.

In Einstein's theory, the quantitative examination of this simple idea meets with difficulties. This is due to the (present) lack of exact solutions which have been clearly identified as describing bounded material systems rotating with respect to each other. Indeed, up to now not even the field of a single, finite, rotating body has been rigorously constructed.[25] However, the latter failure might not be discouraging as a single body would find nothing material to rotate against, except boundary conditions of the type of flat space-time and the like.

This situation recalls the early enthusiasm about Machian properties of the weak field Lense-Thirring metrics describing a mass shell and solid mass sphere, respectively, rotating with respect to asymptotic Minkowski space.[26] This approximate solution exhibits, if only qualitatively, the dragging along of an inertial system by the rotating mass.[27] An improved investigation of this type has been carried through recently by Brill and Cohen.[28] They treat slowly rotating mass configurations as a perturbation of the Schwarzschild geometry rather than of flat space-time. This method admits arbitrarily large masses. In the limit of the mass being equal to the Schwarzschild radius, the dragging along becomes perfect: an inertial system inside the shell rotates with the same angular velocity (with respect to flat space-time) as the shell itself.[29] In the case of two rotating shells the contributions to the dragging along of the inertial system from the shells do not add up in any simple way. Hönl and Soergel-Fabricius avoid the unwanted boundary condition of asymptotical Minkowski space by disturbing the spatially closed Einstein cosmos.[30] They separate out a mass δM and let it rotate around a fixed space-like geodesic. Then, in first approximation, an inertial system is dragged along by the amount $\delta M/M$, where M is the total mass of the model.

This leads to another formulation of the principle of the relativity of rotation: In the rest-system of matter (comoving system), a testparticle must not experience Coriolis or centrifugal forces. The investigations of Hönl and Dehnen indicate that this holds true approximately for the closed Einstein cosmos and all Friedmann models.[31]

The class of solutions of Einstein's equations for which a global rest system of matter exists can be identified with those solutions that admit a normal, time-like congruence of curves for the streamlines of matter. The 1-parameter family of 3-space sections orthogonal to the congruence then defines the global rest system of matter. If such a normal congruence does not exist, a locally comoving observer will notice a rotation of neighboring matter.[32]

The first such exact solution of Einstein's equations depicting a complete homogeneous universe filled with incoherent matter of constant density in which the matter rotates rigidly relative to the local compass of inertia was given by Gödel.[33] He asserted also the existence of a vast class of 'rotating' solutions with incoherent matter of (variable) density and with homogeneous and *finite* 3-space.[34] An example of such a solution with constant density of matter has been displayed by Ozsvath and Schücking.[35] It may be interpreted as a deformed Einstein cosmos. The explicit construction of a rotating solution differing arbitrarily little from a Friedmann model with positive curvature remains to be done.

VI. THE STRONG MACH-EINSTEIN PRINCIPLE

Einstein's 'principle of equivalence' and the appearance of the metric field $g_{\alpha\beta}$ as the fundamental dynamical quantity in his theory removed the basis for a physical or formal distinction between inertial and gravitational force locally. Nonetheless, identifying the knowledge of the inertial behavior of a test particle with the knowledge of the $g_{\alpha\beta}$, Einstein arrived at the formulation of what I shall call the strong Mach-Einstein principle: The whole $g_{\alpha\beta}$-field has to be determined exclusively by the distribution of matter in the universe, i.e. the energy-momentum tensor $T^{\alpha\beta}$ of matter.[36]

Einstein's identification of inertia and metric field is not convincing. The path of a monopole test particle is taken to be a time-like geodesic of the Riemannian manifold either by postulate or by derivation. Such a geometric description does not allow the separating out of inertia and gravitation in a manifestly covariant way. It might be thought of that the mass of the test particle could serve the purpose. However, this is not a geometric concept, the mass being introduced into the geodesic equation by multiplication with a constant factor. It depends on the skill of the particular researcher to define an inertial mass depending on the metric field.[37] As a matter of fact, the concept of mass enters Einstein's theory only via the matter tensor $T^{\alpha\beta}$. Only in special cases has it been possible to define a field mass of gravitation.[38]

Moreover, there are other structures available than just monopole particles. In view of the fact that test particles used for large scale observations are planets, stars, and galaxies most of which possess internal angular momentum, it seems not unreasonable to consider spinning particles also. The general relativistic equations of motion of spinning particles have been studied for a long time.[39] It is known that for a nonvanishing curvature tensor $R^{\alpha}_{\beta\gamma\delta}$ they do not follow geodesic paths and that the knowledge of the metric field alone does not suffice to fix the motion.

It was pointed out a long time ago that the strong Mach-Einstein principle cannot possibly be satisfied by Einstein's theory. Given the $T^{\alpha\beta}$

and regularity conditions for the $g_{\alpha\beta}$ the field equations $R^{\alpha\beta}-\frac{1}{2}Rg^{\alpha\beta}=-\kappa T^{\alpha\beta}$ determine the metric only if additional- and a priori arbitrary-boundary conditions are provided[40] (we come back to this question in Section VIII). Thus, it appears as if the *field* character of Einstein's theory is one of the main obstacles for the formulation of Mach's program.[41]

VII. VACUUM SOLUTIONS AND MACH'S IDEAS

One encounters another, albeit not equivalent, version of the strong Mach-Einstein principle under the terminology of 'degeneration of the metric': In the absence of matter the $g_{\alpha\beta}$ field should vanish.[42] It is understood that this version, though being rather intuitive, has an operational meaning only in the sense of a limiting process. Another formulation quoted in this context is: In the absence of matter, spacetime should necessarily be Minkowskian.[43] It is, however, not easy to understand why this formulation is regarded as an expression of Mach's ideas. Minkowski space-time is as absolute a concept as Newtonian absolute space and absolute time. Furthermore, neither the degeneration of the metric, nor the reduction to the Minkowskian values hold for $T^{\alpha\beta}=0$.

A great many solutions of the vacuum field equations $R^{\alpha\beta}=0$ have been discovered, for example, the static vacuum fields, the Weyl fields, null-fields and their subclass the plane-fronted gravitational waves.[44] The latter class has even a submanifold of geodetically complete solutions,[45] that is, there exist solutions of Einstein's vacuum field equations free of singularities. For them Mach's and Einstein's ideas seem not to apply. The concept of motion of test particles with respect to matter loses its meaning in the absence of material sources represented either by $T^{\alpha\beta}$ or by singularities of the metric. However, the equation of geodesic deviation

$$\delta\ddot{x}^{\alpha}=R^{\alpha}_{\beta\gamma\delta}\,\dot{x}^{\beta}\delta x^{\gamma}\dot{x}^{\delta}$$

describing the relative acceleration of free test particles keeps its meaning even in empty space-time.

One might be tempted to include the 'energy of the gravitational field' among the factors determining the inertia of a body despite the difficulties in Einstein's theory with this concept[46]: a general geometrical quantity describing the localized energy of the gravitational field can be constructed only by introducing additional elements like tetrad fields[47] or another metrical background[48] etc. However, in special situations, for example, if space-time is asymptotically flat (isolated systems), if the Riemannian manifold allows a set of privileged observers (time-like normal congruence) or if it permits certain symmetries one can give a meaning to the concept of energy in the framework of Einstein's theory itself. In general, the use of such a concept for the quantitative formulation of Mach's ideas seems but to increase the difficulties already present.

The investigation of the vacuum field equations shows that the metric field does not owe its existence to matter. Thus, the space-time of General Relativity retains absolute features[49]. Einstein was well aware of this.[50]

VIII. THE STRONG MACH-EINSTEIN PRINCIPLE AND BOUNDARY CONDITIONS

The strong Mach-Einstein principle is invalidated by the arbitrariness of the boundary conditions for the field equations. Several stands have been taken to overcome this situation:

(a) One is content with what I shall call the *weak* Mach-Einstein principle:

The whole $g_{\alpha\beta}$ field has to be determined by the distribution of matter in the universe *and* by suitable boundary conditions. For Einstein, this meant to give up Mach's ideas.[51]

(b) One gets rid of the boundary conditions by changing the global topology of the universe from the Euclidian to a spatially closed one. This point of view has been suggested and favored by Einstein.[51] One now has to pay attention to problems of global topology. It may be conjectured that assumptions about the connectedness at large of the Riemannian manifold will replace the boundary conditions.

(c) One assumes the boundary conditions to be given by averaged out astronomical observations. This is, for example, the position of Gürsey[7] who uses the fact that metrics satisfying both Weyl's postulate and the cosmological principle are conformally flat

$$ds^2 = \lambda(x^0, x^\kappa)\, \eta_{\alpha\beta}\, dx^\alpha\, dx^\beta .$$

By the adoption of the perfect cosmological principle the metric is further restricted to a De Sitter world.[52] Thus, one is able to single out privileged coordinate systems in which the boundary conditions are then formulated

(Section III). The metric tensor is factored into two (density) parts:

$$g_{\alpha\beta} = [(-g)^{1/8}]^2 (-g)^{-1/4} g_{\alpha\beta} \overset{\text{def}}{=} \varphi^2 \gamma_{\alpha\beta}.$$

It is assumed that φ describes the overall cosmological structure and $\gamma_{\alpha\beta}$ (det $\gamma_{\alpha\beta} = 1$) the local gravitational phenomena. The boundary conditions are then, in the privileged frames

$$\varphi \to \lambda, \gamma_{\alpha\beta} \to \eta_{\alpha\beta} \quad \text{for} \quad r \to \infty .$$

Obviously; this discussion limits itself to a small, though physically interesting, class of solutions of Einstein's equations.

(d) One interprets the boundary conditions as junction conditions, having in mind that, observationally, we deal just with part of the universe.[53] Although sufficient for practical purposes, this stand either avoids the cosmological issue of the suggested connection between local inertial behavior and overall matter distribution in the universe or takes a special cosmological model for granted (universe as island of matter in an infinite flat space-time).

(e) The boundary conditions are formulated by mathematical arguments. According to Petrov[54], for example, the metric tensor at the boundary should (1) be of the same algebraic type as elsewhere and (2) allow the maximal isometry group connected with this algebraic type.

IX. MACH'S PRINCIPLE AS A SELECTION RULE

The foregoing discussion shows that Einstein's theory does not incorporate Mach's ideas. Instead, Hönl[55] and Wheeler[56] suggested to supplement Mach's principle as a *separate* cosmological principle. Its task should be one of a *selection rule*, viz. to select a class of physically reasonable models from the large number of solutions already available and still to be found.

It is important in this approach to transform the vagueness of Mach's ideas into a precise mathematical scheme. Then, the selective power of the scheme with regard to Einstein's equations has to be proven. Except for the agreement about the function of Mach's principle and about the role of the (spatial) finiteness of space-time, the mathematical formulations of Hönl and Wheeler have nothing in common. Hence, we will discuss them separately.

A. *Mach's Principle as a Boundary Condition*

Wheeler interprets Mach's principle as a boundary condition for the field equations of gravitation.[57] This approach amounts to spelling out mathematically the combined initial value and boundary value problem for the determination of the $g_{\alpha\beta}$. By this we understand the twofold task of specifying the initial data and of calculating the time development of such data.[58] From Section VII it should be clear that it is not sufficient to give the initial values of the matter distribution alone. Rather, one has to specify (in analogy to the generalized coordinates and moments of mechanics) the 3-geometry on a space-like hypersurface and its rate of change in order to determine the 4-geometry. Wheeler proposes to formulate the initial value problem with the help of the so-called condensed intrasurface variational principle (CIVP). This permits to calculate the 4-geometry provided the regular and closed 3-geometry on a space-like hypersurface at two immediately succeeding instants and the density and flow of mass-energy are given.

As has been pointed out by Wheeler, a great many mathematical questions have yet to be answered in order to put the CIVP to work. However, we are concerned not so much about the mathematical difficulties but with the relation of this approach to Mach's ideas and the formulations discussed before. It seems that the CIVP-formulation of Mach's ideas makes mathematically precise the weak Mach-Einstein principle of Section VIII.

The following example is intended to show how removed from the original ideas of Mach (Section II) this expression of Mach's principle may be. Take the compactified flat 3-space with global topology of a 3-torus ($S_1 \times S_1 \times S_1$) as the 3-geometry to be specified on *both* infinitesimally close space-like hypersurfaces. One expects then the CIVP to give a solution for which the 3-spaces have also *extrinsic* curvature zero.[59] The topology of the 4-geometry would be the product of the real line with the 3-torus ($R \times S_1 \times S_1 \times S_1$), a kind of 4-cylinder. As it is open in the time direction a congruence of straight lines exists for this topology. Therefore, a Minkowskian metric can be laid out on it.[60] This is as valid a solution of Einstein's equations as Minkowski space with global topology $R \times R \times R \times R$. On the other hand, a Riemannian manifold with Minkowskian metric has always been considered to be a good example for an anti-

Machian metric. Thus, the formulation of Mach's principle by Wheeler might not fulfill its role as selection rule for Machian solutions.

B. General Relativistic Dynamics and Mach's Principle

In their suggested formulation of Mach's ideas Hönl and Dehnen[29,61] consider only globally time-orientable solutions of Einstein's equations. Such solutions allow a (global) normed time-like vector field u^α which is interpreted to be the tangent field of a congruence of observers (observer field). The decomposition of the motion of a free test particle with regard to u^α permits then to cast the geodesic equation into the form of Newton's equations and to separate out, though not unambiguously, inertial and gravitational forces. Both forces appear in a way analogous to the Lorentz force in electrodynamics. Formally, u^α plays the role of a vector potential with Dirac gauge. Hönl and Dehnen are able to write down a set of 'Maxwell equations' for the 'gravitational field' (Führungsfeld) $F^{\alpha\beta} = u^{[\beta\alpha]}$.[62] These, however, represent not dynamical equations for u^α but identities modulo Einstein's equations.

The source terms in the gravitational 'Maxwell's equations' $G^{\alpha\beta}{}_{;\beta} = j^\alpha + t^\alpha$ are interpreted as the energy-momentum density of matter and of the gravitational field, respectively. Given a solution $g_{\alpha\beta}$ of Einstein's equations and an arbitrarily chosen but fixed observer field u^α, one can calculate the source terms $j^\alpha + t^\alpha$. One may then *formally* solve for the $G^{\alpha\beta}$ or the 'potential' u^α. If this leads to a regular and *unique* solution[63], this then is interpreted as showing that the inertial behavior of test-particles is determined by the distribution of energy-momentum of matter and the gravitational field (Mach-Hönl-Dehnen principle).

One might feel a little unhappy with this expression of Mach's ideas for the following reasons:

(1) In general, u^α cannot be assumed to represent a *physical* reference system, as no operational recipe for its construction is given (not even in terms of Gedankenexperiment). The motion of an observer, even if considered merely as that of a test particle, cannot be prescribed arbitrarily in Einstein's theory. However, if the Riemannian manifold allows a physical interpretation of u^α e.g. if u^α is tangent to the streamlines of matter, it becomes a legitimate physical reference system.

(2) There is no agreement about the mathematical content of the Mach-Hönl-Dehnen principle.[64]

(3) Its power as a selection rule is rather weak. Among the solutions of Einstein's equations investigated by Hönl and Dehnen only the Minkowski metric, the exterior and interior (fluid sphere) Schwarzschild metric and the weak field Lense-Thirring metric are rated as anti-Machian. On the other side, Hönl and Dehnen conjecture that all spatially closed solutions of Einstein's equations satisfy their formulation of Mach's principle. (Section X.) (Some *open* cosmological solutions do as well.) This puts, for example, the Ozsvath-Schücking rotating solution[35] into the class of Machian solutions, although it does not satisfy the principle of the relativity of rotation (Section V).

X. MACH'S PRINCIPLE AND FINITE UNIVERSE

In both approaches to Mach's ideas described in Section IX it is asserted that Machian solutions of Einstein's equations have closed spatial 3-geometry. In Wheeler's formulation this is fed in by use of the CIVP which is well defined only for finite 3-geometries. However, as stated by Wheeler[65] one may rely as well on a system of differential equations without need for this additional provision.

On the other hand, Hönl and Dehnen take the heuristic point of view that in the absence of boundary conditions at spatial infinity there should be a unique relation between the source terms $j^\alpha + t^\alpha$ and the Führungsfeld $G^{\alpha\beta}$. The example of a compactified Minkowski space (Section IXA) indicates that such an assumption need not be valid. Indeed, Hönl and Dehnen have to exclude explicitly spatially closed solutions which are obtained from open solutions by a suitable change in the connectivity at large.[66]

At present, it is not known in general which global topologies of a space-time manifold are compatible with a given Riemannian metric of signature -2 nor which metrics can be laid out on a Riemannian manifold with given global topology.[67] As long as such questions are not answered it seems scarcely advantageous to use statements about the connectivity at large in the formulation of Mach's ideas.

XI. CONCLUSION

Summarizing this review, it seems appropriate to say that Mach's ideas are neither fundamental for Einstein's theory nor a consequence of it. There is considerable agreement that some solutions of Einstein's equations show Machian properties (Sections IV, V), for example, the Friedmann models [28, 29, 37, 68, 69]. At present, a connection between any of the various formulations of Mach's principle and the (spatial) finiteness of the universe has not been proven. There is no unanimity on the importance of Mach's ideas, the spectrum of opinions ranging from contempt[70] to praise. Some physicists have even conceived of new theories of gravitation with the main aim of incorporating Mach's ideas.[71] Such attempts will be discussed elsewhere.

ACKNOWLEDGEMENTS

The author is pleased to thank Professor P. Havas for a critical reading of the manuscript and for many stimulating discussions on the subject. He is very grateful for the hospitality extended to him by Professor Havas and Temple University.

Temple University, Philadelphia
and
Institut für Theoretische Physik,
Universität Göttingen

Note added in proof: Since completing the manuscript I received notice of another approach toward the formulation of Mach's principle as a selection rule (Section IX).[72] It consists in expressing Einstein's equations in an *integral* form (bitensorial analogue of Kirchhoff's formula for the Poisson equation). While, for an approximative treatment, the metric is represented by a convolution integral of the bitensorial Green's function and the matter tensor, in the general case an integral equation for the $g_{\alpha\beta}$ has to be solved.[73] The question about the uniqueness of the integral formulation remains open. At present, it seems not known which solutions (if any) of Einstein's equations do satisfy this fresh expression of Mach's ideas.

REFERENCES

* This research was supported by the Aerospace Laboratories, Office of Aerospace Research, U.S. Air Force.

[1] I. Newton, *Mathematical Principles of Natural Philosophy*. Revised English translation, University of California Press, Berkeley, 1960, p. 6.

[2] For the much earlier criticism of Newton's absolute space by Huyghens, Leibniz and Berkeley see M. Jammer, *Concepts of Space*, Harper Torch-books, New York 1960, Chap. 4 and J. D. North, *The Measure of the Universe*, Clarendon Press, Oxford 1965, Chap. 16, Sec. 1.

[3] E. Mach, *Die Mechanik in ihrer Entwicklung*, F. A. Brockhaus, Leipzig, 1889, p. 213, 216, 217.

[4] *Ibid.*, p. 218, 485.

[5] *Ibid.*, p. 215/216, 217, 219.

[6] M. Bunge, *Am. J. Phys.* **34** (1966), 588.

[7] F. Gürsey, *Ann. Phys.* (N.Y.) **24** (1963), 211.

[8] F. R. Tangherlini, *Nuovo Cim. Suppl.* **20** (1961), 68.

[9] H. Bondi, *Cosmology*, Cambridge University Press, 1961, p. 28.

[10] V. W. Hughes in *Gravitation and Relativity* (ed. by H. Y. Chiu and W. Hoffmann), W. A. Benjamin, New York 1964, p. 106 (further quoted as G. A. R.).

[10a] M. Strauss, *Synthese* **18** (1968), 251.

[11] F. A. E. Pirani, *Helv. Phys. Acta Suppl.* **4** (1956), 199.

[12] B. Bertotti in *Evidence for Gravitational Theories* (ed. by C. Møller), Academic Press, New York and London 1962, p. 179 (further quoted as E.G.T.).

[13] H. Y. Chiu, W. F. Hoffman, in G.A.R., p. xx. Mach himself seems to have thought that the influence of nearby matter cancels out. Cf. ref. 3, p. 219.

[14] P. Havas, *Rev. Mod. Phys.* **36** (1964), 938. See also P. Havas, 'Foundation Problems in General Relativity' in *Delaware Seminar in the Foundations of Physics* (ed. by M. Bunge), Springer-Verlag, Berlin, 1967, p. 124.

[15] A. Trautman, in 'Lectures on General Relativity', *Brandeis Summer Institute in Theoretical Physics* 1964, vol. I, Englewood Cliffs, N.J., 1965, p. 104, 115 (further quoted as B.G.R.).

[16] B. Bertotti, D. Brill, R. Krotkov, in *Gravitation* (ed. by L. Witten), John Wiley, New York, 1962, p. 4 (further quoted as G.I.C.R.).

[17] A. Trautman, in B.G.R. ref. 15, p. 137.

[18] A. D. Fokker, *Proc. K. Akad. Wet. Amsterdam* **23** (1920), No. 5.

[19] For extensive discussions of Einstein's interpretation of Mach's ideas compare: H. Hönl, *Physikertagung Wien* 1961, p. 88–105, Physik Verlag, Mosbach, 1962; *Wissenschaftl. Zeitschr. d. Friedr. Schiller Univ. Jena., Mathem. Naturw. R.* **15**, 25–36 (1966); *Einstein Symposium 1965*, p. 239–276, Akademie-Verlag, Berlin, 1966.

[20] A. Einstein, *The Meaning of Relativity*, Princeton University Press, 1955, p. 56.

[21] *Ibid.*, p. 99.

[22] *Ibid.*, p. 100.

[23] There seems to exist disagreement about the magnitude of the first effect, however. Compare H. Dicke, 'The Many Faces of Mach', in G.A.R. p. 123–124 and H. Dehnen, H. Hönl and K. Westpfahl, *Ann. Physik* **6** (1960), 388.

[24] Reference 1, p. 10–11.

[25] There are indications that the Kerr metric (R. P. Kerr, *Phys. Rev. Letters* **11** (1963), 237) describes the exterior field of a finite, axially symmetric, rotating body. However, the identification with a definite material system of a matching interior solution still seems to be open. Compare W. C. Hernandez, *Phys. Rev.* **159** (1967), 1070; **167** (1968), 1180 and V. De La Cruz and W. Israel, *Phys. Rev.* **170** (1968), 1187.

[26] H. Thirring, *Phys. Z.* **19** (1918), 33; **22** (1921), 29; H. Thirring and J. Lense, *Phys. Z.* **19** (1918), 156.
[27] See also the note of C. Cattaneo, *Ac. Naz. Lincei Rend., Sc. Fis. math.* **32** (1962), 346, on the rigorous gravitational field of a rotating mass shell. However, we cannot accept his conclusions concerning Mach's principle.
[28] D. Brill and J. Cohen, *Phys. Rev.* **143** (1966), 1012; D. Brill and J. Cohen, *Nuovo Cim.* **56B** (1968), 209.
[29] This result was also obtained for a rotating fluid sphere by H. Hönl and H. Dehnen, *Ann. Physik* **14** (1964), 293.
[30] H. Hönl and Ch. Soergel-Fabricius, *Z. Physik* **163** (1961), 571.
[31] H. Hönl and H. Dehnen, *Z. Physik* **166** (1962), 544. H. Dehnen, *Z. Physik* **166** (1962), 599.
[32] J. L. Synge, *Proc. Lond. Math. Soc.* **43** (1937), 376.
[33] K. Gödel, *Rev. Mod. Phys.* **31** (1949), 447.
[34] K. Gödel, *Proceedings of the International Congress of Mathematicians 1950*, vol. I p. 180, American Mathematical Soc. Providence 1952.
[35] I. Ozsvath and E. Schücking, *Nature* **193** (1962), 1168.
[36] Reference 20, p. 103 and Letter of Einstein to Mach (1913) publ. by H. Hönl, *Phys. Bl.* **11** (1960), 571.
[37] See, for example N. Rosen, *Ann. Phys.* (N.Y.) **35** (1965), 426.
[38] For *static fields* compare J. Ehlers and W. Kundt, 'Exact Solutions of the Gravitational Field Equations', *G.I.C.R.* (ref. 16), p. 68–69.
85. *Mass Aspect:* H. Bondi, M.G.J. van der Burg, and A. W. K. Metzner, *Proc. Roy. Soc.* (London) **A269** (1962), 21. For the mass of a gravitational wave pulse see D. R. Brill, *Ann. Phys.* (N.Y.) **7** (1959), 466.
[39] M. Mathisson, *Acta Phys. Polon.* **6** (1937), 163, 218; A. Papapetrou, *Proc. Roy. Soc.* (London) **A209** (1952), 248.
[40] G. Beck, *Z. Physik* **33** (1925), 713.
[41] Compare Einstein's autobiographical notes in *Albert Einstein: Philosopher-Scientist* (ed. by P. A. Schilpp), Evanston, Ill., 1949, p. 29.
[42] A. Einstein, *Sitzber. Preuss. Akad. Wiss.* **142** (1917).
[43] Comp. e.g. W. Thirring, *Fortschr. Physik* **7** (1959), 80.
[44] J. Ehlers and W. Kundt, ref. 38 p. 49–99; The plane-fronted waves are the vacuum null fields which possess a normal, non-shearing and non-expanding ray congruence.
[45] A Riemannian manifold is called geodetically complete if every geodesic has infinite length in both directions. For a discussion of various concepts of completeness compare W. Kundt, *Z. Physik* **172** (1963), 488, and ref. 60 (Fierz and Jost).
[46] For a review see A. Trautman, *Conservation Laws in General Relativity* in *GICR*, p. 188. Compare also C. Cattaneo, *Ann. Inst. Henri Poincaré* **4** (1966), 1.
[47] C. Møller, *Kgl. Danske Videnskab. Selskab, Mat.-Fys. Skrifter* **1** (1961), No. 10; *Mat.-Fys. Medd.* **34** (1964), No. 3.
[48] N. Rosen, *Phys. Rev.* **57** (1940), 147, 150.
M. Kohler, *Z. Physik* **131** (1952), 571; **134** (1953), 286, 306.
[49] For a discussion see A. Grünbaum, *Philosophical Problems of Space and Time*, A. Knopf, New York 1963, Chap. 14; and *Phil. Review* **66** (1957), 525.
[50] A. Einstein, Preface to M. Jammer, ref. 2, p. XV.
[51] See ref. 42 or ref. 20 p. 108.
[52] *Weyl's postulate:* The particles representing the nebulae lie in space-time on a bundle of geodesics diverging from a point in the past. *Cosmological Principle:* The

universe presents the same aspect from every point in space. *Perfect Cosmological Principle:* The universe presents the same aspect from every point in space at any time.

[53] O. Klein in *Recent Developments in General Relativity*, Pergamon Press, Oxford, 1962, p. 293; H. Hönl and H. Dehnen, *Ann. Physik* **11** (1963), 209.

[54] A. S. Petrov, *Einstein-Räume*, Akademie Verlag, Berlin, 1964, p. 357.

[55] H. Hönl, *Z. Naturforsch.* **8a** (1953), 2.

[56] J. A. Wheeler in *La structure et l'évolution de l'univers*, R. Stoops, Bruxelles, 1959, p. 49–51.

[57] J. A. Wheeler, 'Mach's principle as boundary condition for Einstein's equations', in *Boulder Lectures in Theoretical Physics* 1962, Interscience, 1963, p. 528; *GAR* p. 303; in *Conférence Internationale sur les théories relativistes de la gravitation*, Warsaw 1962, Gauthier–Villars, Paris, 1964, p. 223.

[58] See Y. Bruhat, 'The Cauchy Problem in Gravitation', in GICR (ref. 16) p. 130.

[59] The existence of the 4-geometry described is proven by theorem 1 of R. P. Geroch, *J. Math. Phys.* **8** (1967), 782.

[60] M. Fierz and R. Jost, *Helv. Phys. Acta* **38** (1965), 138; N. Steenrod, *The Topology of Fibre Bundles*, Princeton University Press, 1967, p. 207.

[61] H. Hönl and H. Dehnen, *Ann. Physik* **11** (1963), 201; *Z. Physik* **191** (1966), 313; *Z. Physik* **206** (1967), 492.

[62] For the detailed description taking into account the distinction between field strength $F^{\alpha\beta}$ and field intensities $G^{\alpha\beta}$ and establishing constitutive equations see ref. 61. (*Z. Physik* **191**). One gets $G_{\alpha\beta} = (2\kappa)^{-1} F_{\alpha\beta} - 3(2\kappa)^{-1} u_{[\alpha}F_{\beta]\gamma}u^{\gamma}$. $(A_{[\alpha\beta]} = A_{\alpha\beta} - A_{\beta\alpha})$.

[63] Which coincides, of course, with $G^{\alpha\beta}$ directly calculated from the chosen u^{α}.

[64] See J. Ehlers and E. Schücking, *Z. Physik* **206** (1967), 483.

[65] See ref. 57 (GAR) p. 330.

[66] Ref. 61 (*Z. Physik* **206**), p. 501, footnote.

[67] For a discussion of the problem cf. W. Rinow, *Die innere Geometrie der metrischen Räume*, Berlin, 1961. Some general results for compact differentiable manifolds are found in ref. 60 (Steenrod) while results for homogeneous spaces of constant curvature (and arbitrary signature) are given by J. A. Wolf, *Spaces of Constant Curvature*, McGraw-Hill, New York, 1967. A discussion of flat Lorentz 3-manifolds is found in L. Auslander and L. Markus, *Amer. Math. Soc. Memoir No. 30*, Providence 1959.

[68] P. Szekeres, *J. Math. Phys.* **6** (1965) 1387.

[69] J. M. Cohen, 'Gravitational Collapse of Rotating Bodies', Institute for Space Studies, Goddard Space Flight Center, preprint March 1968.

[70] Ya. B. Zeldovich, in *Advances in Astronomy and Astrophysics*, vol. III, Academic Press, New York, 1965, p. 368.

[71] See for example D. W. Schiama, *Monthly Notices Roy. Astron. Soc.* **113** (1953) 34 – *The Unity of the Universe*, Faber and Faber, London, 1959, Chap. 9. W. Davidson, *Monthly Notices Roy. Astron. Soc.* **117** (1957), 212. R. Dicke, in EGT (ref. 12) p. 36.

[72] B. L. Altshuler, *JETP* **24** (1967), 766; D. Lynden-Bell, *Monthly Notices Roy. Astron. Soc.* **135** (1967), 413.

[73] D. W. Sciama and P. C. Waylen, 'A Generally Covariant Formulation of Einstein's Field Equations', preprint, University of Cambridge, England (1968).

APPENDICES

APPENDIX A

PHILIPP FRANK

THE IMPORTANCE OF ERNST MACH'S PHILOSOPHY OF SCIENCE FOR OUR TIMES*

There is something remarkable about the teachings of Mach. Philosophers often ridicule or disdainfully reject them as the work of a physicist dabbling in philosophy; physicists often deplore them as aberrations from the right path of respectable, realistic natural science. Yet neither physicists and philosophers, nor historians and sociologists, nor many others, can get rid of Mach. Some attack him passionately; others extol him with fervor. There is something fascinating about his simple, straightforward teachings. In spite of their simplicity they are stimulating and provocative. There are indeed but few thinkers who can induce such sharp differences of opinion, who are so inspiring to some and so utterly repugnant to others. What is there in these doctrines which makes it impossible for anyone, whatever his views may be, to avoid adopting some definite attitude towards them?

This is what I should like to discuss in the present paper. I have formed a fairly definite opinion about the position which Mach occupies in the intellectual life of our times, and this position, I believe, will explain why the battle rages about him so furiously. It is not a question here of the details of Mach's teaching, often individually and historically conditioned, but rather of their nucleus, which is just the focus of the struggle. I will not speak therefore about the general attitude of Mach towards the psycho-physical problem, nor about his separate contributions to physics and psychology, but only about his conception of the tasks and possible aims of exact science.

In recent years, among creatively active physicists and mathematicians, there has become noticeable a reaction against the conceptions of Mach. When one of the most outstanding theoretical physicists of our time, Max Planck[1], and one of the foremost living geometricians, E. Study[2], characterize these conceptions as being partly misleading, partly incapable of being carried out, and partly actually harmful for science, the fact provides food for thought and cannot be lightly brushed aside.

What an investigator with the markedly constructive talents of Planck dislikes above all in the views of Mach is his judgment of values. For the investigator, every new theory which is supported by experiment is a piece of newly discovered reality. According to Mach, on the other hand, physics is nothing but a collection of statements about the connections among sense perceptions, and theories are nothing but economical means of expression for summarizing these connections.

"The aim of natural science," says Mach, "is to obtain connections among phenomena. Theories, however, are like withered leaves, which drop off after having enabled the organism of science to breathe for a time."[3] This phenomenalistic conception, as it is called, was already familiar to Goethe. In his posthumous 'Maxims and Reflections' he says: "Hypotheses are the scaffolds which are erected in front of a building and removed when the building is completed. They are indispensable to the worker; but he must not mistake the scaffolding for the building." And still more drastically: "The constancy of phenomena alone is important; what we think about them is quite immaterial."

It will be said, however, that Goethe was not really a good physicist, and that we can see in his case an example of how such basic principles hinder the spirit of research. Thus Planck says: "When the great masters of exact investigation of nature gave their ideas to science, when Nicholas Copernicus removed the earth from the center of the universe, when Johannes Kepler formulated the laws named after him, when Isaac Newton discovered gravitation ... – the series could be long continued – surely, economical points of view were the very last thing to steel these men in their struggle against traditional opinions and dominating authorities. No, it was their unshakable belief – whether resting on an artistic, or on a religious basis – in the reality of their world picture. In view of these certainly incontestable facts, one cannot reject the surmise that, if the Mach principle of economy were really to be put at the center of the theory of knowledge, the trains of thought of such leading spirits would be disturbed, the flight of their imagination crippled, and consequently the progress of science perhaps fatefully hindered."[4]

That these fears in such generality are groundless can be readily seen if one recalls the views of one of the greatest theoretical physicists of the nineteenth century, James Clerk Maxwell, on the nature of physical theories. One need only read the introduction to his essay on Faraday's lines of force (1855)[5] to be convinced that he was completely an adheren of the phenomenalistic standpoint. Yet one cannot say of him in any way that adherence crippled the flight of his imagination; indeed, quite the opposite. The conception of the relative worthlessness of the theory in comparison to the phenomenon gives to the theorizing of such an investigator something especially free and imaginative.

I am willing to concede that the phenomenalistic doctrine is pleasing to those whose work in physics is descriptive rather than constructive. Many such people, who are capable of describing very neatly definite – even if very special – phenomena, may regard themselves because of this doctrine as being sublimely superior to the imaginative, creative spirit, whose works, after all, are only phantoms and "withered leaves". I do not believe, however, that for people with such character the philosophy of Mach has crippled the imagination. Rather, it is the case of an imagination crippled by nature which uses the Mach teachings as a beautiful cloak to cover its deformity. It may have been cases like these which cause Planck, at the end of the lecture cited above, to hurl at the preachers of the phenomenalistic doctrines the biblical words: "By their fruits ye shall know them."

Concerning this criterion of the fruits I shall have more to say. First, in connection with the same biblical allusion, I should like to introduce a quotation from Pierre Duhem about the value of physical theories. Duhem was the most outstanding representative in France of ideas similar to those of Mach. He said: "By the fruit one judges the tree; the tree of science grows exceedingly slowly; centuries elapse before one can pluck the ripe fruits; even today it is hardly possible for us to shell and appraise the kernel of the teachings that blossomed in the seventeenth century. He who sows cannot therefore judge the worth of the corn. He must have faith in the fruitfulness of the seed in order that he may follow untiringly his chosen furrow when he casts his ideas to the four winds of heaven."[6]

These remarks of the greatest and most accurate student of the history of physics are perhaps the best answer to the opinion expressed by Planck "that even our present world picture, although it shows the most varied colors according to the individuality of the investigator, nevertheless possesses certain features which can never be obliterated by any revolution, either in nature or in the human mind".[7] These enduring features arise, according to Mach, from the fact that all possible theories must give the same connection between phenomena; this very fact guarantees a certain constancy. The known connections among phenomena form a network; the theory seeks to pass a continuous surface through the knots and threads of the net. Naturally, the smaller the meshes, the more closely is the surface fixed by the net. Hence, as our experience progresses the surface is permitted less and less play, without ever being unequivocally determined by the net.

Since the basic principles of Mach would do nothing but harm to physics, it is fortunate for physics, according to Planck and Study, that these principles have never been thoroughly applied by their adherents, even if it is a gloomy sign for the principles themselves. Thus Study says about positivism, as he calls the doctrines of Mach: "We regard this principle as a perfect utopianism. The possibility of its existence is based entirely on the fact that it is disavowed by its own followers at every step. Up to now there has never been any serious attempt to apply it consistently." "We are dealing here with a question of principle and must therefore distinguish between the theory of positivism and the practice of the (fortunately for themselves) thoroughly inconsistent positivists."[8] Planck says similarly: "We then attain a more realistic mode of expression, ... which is actually the one always used by physicists when they speak in the language of *their* science."[9]

With biting sarcasm, Study says: "In numerous cases the hypotheses which are basely denounced at the official reception (why not atomistics, too?) are admitted, under a different name and through a back door especially arranged for this, into the sanctuary of science. Such names and corresponding motivations are by no means few. Without any effort the writer collected a full dozen of them: 'most complete and simplest description' (Kirchhoff), ... 'subjective means of research', 'requirement of conceivability of facts', 'restriction of possibilities', 'restriction of expectation', 'result of analytical investigation', 'economy of thought', 'biological advantage' (all of these employed by E. Mach)."[10]

With equal mockery, Planck remarks: "I should not be at all surprised if a member of the Mach school were to come out some day with the great discovery, that ... the reality of atoms is just what is required by scientific economy."[11]

Other authors, too, point out the glaring contradiction that exists among the admirers of Mach between theory and practice. A peculiar theory of the nature of physical theories is set up, but as soon as physics really begins the positivist behaves practically like any other physicist. A follower of Mach can proclaim that physics has to do only with relations among sense perceptions, but the preacher of this doctrine speaks as a physicist exactly like any one else, about matter and energy, and even about atoms and electrons.

However, it is just this apparently so palpable contradiction that can lead to the understanding of the permanent nucleus of Mach's teachings. Let us listen once more to Study: "The whole situation is a striking reminder of Kronecker's proposal to abolish the irrational numbers and to reduce mathematics to statements about integers; in this case, too, the suggestion has remained programmatic, and for the same good reasons."[12] The analogy, as I see it, is a very appropriate one. But I should like to give it a different interpretation from that of Study. It is obviously pointless actually to express all theorems of mathematics as theorems about integers. In principle, however, it is highly enlightening to know that all theorems about irrational numbers, and hence also all theorems about limiting values, *could* be expressed as theorems about integers. Once this possibility has been substantiated, the whole of analysis can proceed to develop as usual. But now when a theorem about derivatives is set up and somebody begins to subtilize about it, asking whether this theorem is really in agreement with the 'nature' of the differential and going into profound and skeptical deliberations concerning this 'nature', he can be told quite simply: "I could express this theorem, if I took enough time, as a theorem about integers; the nature of this theorem is hence no more and no less mysterious than that of the natural numbers."

The situation is quite similar with respect to Mach's physical theory of knowledge. It is not a question of actually expressing all physical statements as statements about relations among sense perceptions. It is important, however, to establish the principle that only those statements have a real meaning which *could* in principle be expressed as statements about the relations among our perceptions. To express the law of conservation of energy or the law of equipartition of energy among all the degrees of freedom as statements about relations among perceptions is just as laborious, but also just as superfluous, as to express the theorem that the derivative of the sine is the cosine as a statement about integers. In principle, however, both are certainly possible.

For the inner working of physics it is in most cases practically immaterial whether one has adopted Mach's standpoint or not. Similarly, in Kronecker's lectures on integral calculus there is nothing to be found that differs essentially from the presentation of other mathematicians.

Wherein, then, lies the value of the doctrines of Mach for physics? My view is that their main value is not that they help the physicist to go forward in his physical work, but rather that they provide the means for defending the edifice of physics against attacks from outside.

One who examines dispassionately the concepts which today are at the basis of the system of hypotheses of physics will hardly be able to assert seriously that the atom, the electron, and the action quantum form really satisfactory ultimate building blocks. Every thinker who is somewhat inclined towards logical thoroughness can find many obscurities in these concepts. Into these nebulosities boring doubt can penetrate and try to shake the whole system of physics as the foundation of our scientific world picture. Here Mach steps in and says: "All these concepts are only auxiliary concepts. What is significant is the connection among phenomena. Atoms, electrons, and quanta are only links to represent a connected system of science; they make it possible to derive logically the immeasurable system of connected phenomena from a few abstract principles. But these abstract principles are then only the means to an economical representation. They are not the epistemological basis. The reality of physics can never be shaken by any criticism of the auxiliary concepts." The work of Mach is therefore not essentially destructive, as it is often represented; positivism is not, as Study calls it, a "negativism", but on the contrary is an attempt to create an unassailable position for physics. As a matter of fact, Planck, too, recognizes this when he says: "To it [the positivism of Mach] belongs in full measure the credit for having found again, in the face of threatening skepticism, the only legitimate starting point of all investigation of nature, the sense perceptions."[13]

That Planck condemns so sharply the conception of Mach seems to me therefore to arise from the fact that he considers it only from the viewpoint of its application within physics. It must be said, however, that, even when regarded from this standpoint, the phenomenalistic conception has already accomplished something and is perhaps capable of accomplishing still more. In the boundary regions of physics, where general concepts like space, time, and motion play a role, the epistemological position one takes is no longer entirely immaterial. Indeed, it is universally known today that the Einstein general theory of relativity and gravitation grew immediately out of the positivistic doctrine of space and motion, as Einstein himself has discussed in detail in his reference to Mach.[14]

On the whole, however, I will gladly concede to Planck and Study that positivism itself has not done much in clearing up individual questions of physics. But from this it does not follow that positivism is, in general, of no value. The 'fruits' of Mach's teachings are indeed not purely physical in character. It will be recalled that in recent years an attempt was made to take advantage of the criticism of the basic physical concepts in order to proclaim the bankruptcy of the scientific world picture. Bearing this in mind, one must ascribe a high value to the effort of Mach to make physics independent of every metaphysical opinion.

H. Poincaré says: "At first glance it appears to us that theories last only a day and that ruins heap up on ruins. ... If we examine the matter more closely, however, we find that what decays is those theories which claim to teach us what things *are*. But there is something in them which endures. If one of them has revealed to us a true *relation*, this relation has been acquired for all time. We shall find it again under a new cloak in the other theories which will reign successively in its place."[15]

In a very determined manner, the French philosopher Abel Rey emphasizes the importance of preserving the edifice of physical ideas for the general intellectual life. He says: "If those sciences which have had an essentially emancipating effect in history go down in a crisis which leaves them only with the significance of technically useful collections but robs them of every value in connection with the cognition of nature, this must bring about a complete revolution in the logical art. The emancipation of the mind as we owe it to physics is a most fatally erroneous idea. One must introduce another way, and restore to a subjective intuition, to a mystical sense of reality – in short, to the mystery – everything that we believed had been taken away from it. If, on the contrary, it turns out that there is no justification for regarding this crisis as necessary and incurable, then the rational and positive method remains the best nurse of the human spirit."[16]

Here we have a very clear exposition of the dangers that would arise for the whole world conception from a physics which had no other epistemological foundations than those auxiliary concepts which are so much exposed to criticism.

Mach himself saw the real value of his theories in the fact that they permitted setting up a connection, as free from contradictions as possible, between physics on the one hand and physiology and psychology on the other. He who still doubts this has only to read the general sections of the *Analysis of Sensations*.[17] Here it is stressed again and again that one must exert oneself to develop physics while using concepts which will not have to be given up in a transition to a neighboring branch of science.

From this effort of Mach to employ only concepts which will not lose their usefulness outside of physics we can understand his opposition to atomistics, which in particular has caused many physicists to turn against him. It is true that atomistics, when applied to physiological and psychological problems, easily leads into a blind alley. Such questions arise as: "How can a brain atom think?" "How can an atom perceive green, since, after all, it is itself again only a miniature picture of a macroscopic body composed of perceptions?"

I will not deny that Mach allowed himself to be misled by this argument into attacking the use of atomistics in physics more sharply than can be justified. After all, the usefulness of the atomic theories in this limited realm is certainly indisputable. His followers, as is generally the case, often saw in this weakness of the master his greatest stringth and wished to banish the atom entirely from physics. I believe that one can completely free the nucleus of Mach's teachings from this historically and individually conditioned aversion to atomistics. The atoms are auxiliary conceptions just like others which can be employed advantageously in a limited domain. They are not suitable for an epistemological foundation. Once we have adopted this point of view, we are all the freer in employing the concept of atoms wherever it is admissible. I believe that even Planck would not object so much to the kernel shelled in this way. It is then no longer so very strange when one declares the atoms, if not their reality, to be a requirement of economy. They can be the simplest means of representing physical laws without thereby being suitable to form an epistemological foundation.

In general, phenomenalism will neither particularly further nor hinder the physicist in his field of work. Thus Maxwell, who doubtless thought positivistically, wrote the work that laid the foundation for the molecular theory of gases. The phenomenalistic conception becomes a danger only in those cases where the requirement of economy is not realized with equal intensity. The most noteworthy historical example is perhaps Goethe's doctrine of colors. However, if one wishes to pass judgment on a person of such strong individuality one must not forget, as A. Stöhr quite correctly points out, that the requirement of economy may mean something quite different for every individual.[18] For one it signifies a minimum of hypotheses, for another, say, a minimum of different kinds of energy. The former is the case for the extreme phenomenalist Goethe; the latter for the pure mechanists.

It will perhaps be instructive, as a comparative case, to recall a theoretical physicist who, as an immediate student of Mach, really tried to construct a system of physics and chemistry in which no hypothetical corpuscles, whether atoms or electrons, occur, and which embraces all the phenomena known at present. It cannot be denied that Gustav Jaumann in numerous works has undertaken this task with great constructive force.[19] I do not believe, however, that the result has turned out to be really in the spirit of Mach's teachings. To be sure, it corresponds to the surface requirement that all atomistics be omitted, but it hardly corresponds to the requirement of economy. A large number of constants are employed about which the theory can make no prediction. The Jaumann system makes it possible only in a very limited degree to derive phenomena, also with regard to numerical values, from a small number of hypotheses. To show the independence of physical research and the epistemological basis, one can mention that the most energetic attempt to refute the corpuscular theory of electricity, that of F. Ehrenhaft, has no connection whatsoever with philosophical dogmas of any kind.

I believe that I have now made clear to some extent the significance of Mach. In order fully to survey his position in the intellectual life of our times, however, we must find a more remote standpoint, in order to obtain a better view.

If we read the most important work of Mach, his *Mechanics*,[20] we shall find that in no section does he give us so deep an insight into his innermost thoughts and intellectual inclinations as in the wonderful chapter on theological, animistic, and mystical points of view in mechanics. A wind of refreshing coolness blows from these statements. What other authors have treated with vehement blustering, often with a quiet announcement of a small *auto-da-fé* for the opponent, we see here discussed in a genuinely scientific spirit. Yet throughout the whole chapter there quivers an undertone of suppressed excitement. One encounters there that state of being drunk with soberness which has been attributed to the Age of Enlightenment. Indeed, Mach descries in this age his spiritual home. In the chapter referred to, he says: "For the first time, in the literature of the eighteenth century enlightenment appears to be gaining a broader base. Humanistic, philosophical, historical, and natural science come into contact at this time and encourage one another to freer thought. Everybody who has experienced this upsoaring and emancipation, even if only in part, through the literature, will feel a lifelong melancholy nostalgia for the eighteenth century."

The personal acquaintances of Mach are aware that he was an ardent admirer and reader of Voltaire. One of his former assistants, Professor George Pick informed me that Mach most emphatically condemned Lessing's attacks on Voltaire. It is known, too, that Josef Popper, of whom Mach says that for a long time he was the only man with whom he could talk of his physical and epistemological opinions without precipitating a conflict, wrote a whole book devoted to the defense, indeed the glorification, of Voltaire.

In my opinion, Mach was led in this predilection by a correct estimate of himself. We can understand the role which as philosopher he plays in the intellectual life of the present if we look upon his teachings as the enlightenment philosophy appropriate to our time.

Since this conception may easily be misunderstood, I must discuss it more fully. First of all, the word 'enlightenment' has acquired such a bad connotation that perhaps many will see in this description a disparagement of Mach. We must therefore try to make clear something about the nature of this enlightenment and the reasons for its subsequent neglect.

The first period of enlightenment in modern times began with the downfall of the Ptolemaic world system. Copernicus tried to represent his system with the help of the concepts of the Aristotelian scholastic philosophy. If, however, we read the dialogues of Galileo about the two world systems, we encounter a very different tone. Here the basic concepts of Aristotelian physics were taken up and examined. In the teachings of Aristotle and his school, concepts like *light* and *heavy*, *above* and *below*, *natural* and *forced* motion, which were usable only for a very limited domain of experiences, were made the basis of all theoretical physics. Galileo showed that it was just this use of concepts outside of their natural realm of validity that prevented the followers of Aristotle from understanding modern physics. I do not mean by this to belittle Aristotelian physics, which was an outstanding contribution for its time; I only wish to show that what was enlightening in Galileo's writings was his setting a limit to the misuse of auxiliary concepts. And it is this protest against the misuse of merely auxiliary concepts in general philosophical proofs that I consider to be an essential characteristic of enlightenment. Every period of physics has its auxiliary concepts, and every succeeding period misuses them. Hence in every period a new enlightenment is required in order to abolish this misuse. When Sir Isaac Newton and his contemporaries made the concepts of absolute space and time the basis of mechanics, they were able to represent a large domain of physics properly and without contradictions. It does not follow, however, that these concepts form a basis of mechanics that is satisfactory from the standpoint of the theory of knowledge. When Mach criticized the foundations of Newtonian mechanics and tried to eliminate absolute space from it, he was the direct continuer of the work of Galileo. For in absolute space we still had a remnant of Aristotelian physics. And when Einstein joined Mach and in his general theory of relativity really erected an edifice of mechanics in which space and time properly speaking no longer occurred, but only the coincidence of phenomena, the elimination demanded by Mach of the auxiliary concepts of space and time – useful only in a limited domain – was now completed. In Einstein we see the first thinker to found a physics entirely free of Aristotelianism.

In the Age of Enlightenment proper I also see a struggle against the misuse of auxiliary concepts. If we leave out of the discussion the political and social aspects, then, theoretically considered, the criticism at that time was directed against the fact that theological concepts which were formed for dealing with certain psychic experiences of human beings were made the foundation of all science throughout the Middle Ages and even at the beginning of the modern era. These concepts, no matter how appropriate they may be to restore hope and faith to the struggling human soul, are nevertheless only auxiliary concepts limited to this domain and are not suitable to be the epistemological foundation of our knowledge of nature. This critical point of view emerged with great energy at that time. Today even theologians have adopted the view that the Bible is not a scientific textbook. Indeed, many Protestant theologians, proceeding still further in the direction of enlightenment, now teach that all theological truths are only statements about inner experiences.

The natural science of the Enlightenment also needed auxiliary concepts for its development. In this way the concepts of matter and atom began to play a decisive role. All at once these auxiliary concepts were being applied to everything in the world; materialism, as it is called, was born. The fact that matter, too, was only an auxiliary concept was forgotten, and people began to regard it as the essence of the world. Soon criticism of this view set in. But although this criticism of the misuse of auxiliary concepts usually serves only scientific progress, here it had an additional effect. Since the ideas of the Age of Enlightenment were not pleasing to the ruling powers, the criticism of the misuse of enlightenment was used to discredit enlightenment itself. Because the rationalists misused auxiliary concepts, it was said of them that their protest against the theological world picture was unjustified. This view is, of course, logically not tenable, for the fact is that their criticism did not go far enough. However, there are always thinkers who are so constituted that their thinking ultimately leads to the conclusions required by the ruling powers. An attempt was made to overthrow enlightenment by skepticism. Very appropriately Nietzsche says, of the part taken in this work by certain philosophers: "The philosopher against his rival, e.g., science: now he becomes a skeptic; now he reserves for himself a form of cognition which he denies to the scientist; now he goes hand in hand with the priest so as not to arouse the suspicion of atheism, materialism; he regards an attack on himself as an attack on morality, virtue, religion, order – he knows how to discredit his opponent as a 'seducer' and an 'underminer'; now he goes hand in hand with the authorities."[21]

In reality, however, only that part of enlightenment was refuted which was not enlightenment. Nevertheless, because of the weight of authority of those involved, this disparagement of the great achievements of the eighteenth century has had considerable influence. There is perhaps no one among us who did not acquire a prejudice against enlightenment at school during his youth.

I admit, of course, that the great spirits of the Enlightenment, Voltaire, d'Alembert, etc., were imitated by many shallow writers who diluted their criticism more and more and descended to intolerable banality, ending up by continuing themselves the misuse of the auxiliary concepts.

I admit, too, that this shallowness belongs to the essence of the Enlightenment. Once the misuse of the old concepts has been exposed, there is not much of an original nature left to say. The temptation to dull triviality is strong and the number of those who have fallen victim to it is great. All this, to be sure, provides no evidence whatsoever against the enlightenment philosophy.

Having freed himself once and for all from the fears of the usual stigma of heresy, one will say that the task of our age is not to fight against the enlightenment of the eighteenth century but rather to continue its work. Since that time there has occurred so much exaggerated application of entirely new auxiliary concepts which are useful in limited domains that there is plenty of new work to be done.

To this work Mach dedicated himself. He approved enthusiastically of the 18th-century enlightenment. This does not mean, however, that he began to idolize the 18th-century concepts like materialism. Rather, in him lived the spirit of those great men; it drove him on to protest against the misused concepts of his time just as they had fought those of their time. Among the concepts that he fought there happened to be many of the favorite concepts of 18th-century enlightenment.

This is what I mean when I call Mach the representative of the enlightenment philosophy of our period. Since his youth was lived during the time of materialism, it is no wonder that so many of his works are devoted to the struggle against mechanistic physics and atomistics.

If we accept Mach's attitude as that of an enlightenment philosopher it will be easier for us to understand many features of his teachings and many of their effects. In the first place, there is their strongly suggestive influence, one might even say their virulence, which in spite of many contemptuous judgments by professional philosophers, commands attention. Study calls the positivism of Mach "a still completely unsatisfied existence, a kind of philosophical beast of prey, hungry for a victim".[22] As in the case of the philosophers of the enlightenment, so also in the case of Mach, it developed that the disciples and followers displayed an excessive tendency towards shallowness. Furthermore, to Planck's criterion of the fruits the present opinion provides an answer: the fruits of Mach's teachings are not the writings of his physical and philosophical followers, but rather the enlightenment of minds brought about by them – a fact which even Planck recognizes. By this I do not mean to assert that Mach has no importance in other respects. I do think, however, that this is the best summary of his position in the general intellectual life of our times.

In this opinion I am strengthened also by the striking agreement of his views with those of a thinker for whom he cannot have had any great sympathy, Friedrich Nietzsche. This agreement was first pointed out by Kleinpeter.[23] The more one delves into the posthumous works of Nietzsche, the more clearly one observes the agreement, particularly in the basic ideas related to the theory of knowledge. Now Nietzsche is the other great enlightenment philosopher of the end of the nineteenth century. The harmony of his epistemological views with those of Mach, who had gone through an entirely different course of instruction and possessed an entirely different temperament and entirely different ethical ideals, seems to me to be evidence for the fact that such views must have penetrated to the enlightened minds of that time.

That great master of language, Nietzsche, formulated these ideas with extraordinary force and impressiveness when he said: "I behold with amazement that science today has resigned itself to being relegated to an apparent world; a true world, be it what it may: in any case, we have no organ of cognition for it". Here one might ask: by what organ of cognition is one led to assume this antithesis?... From the fact that a world which is accessible to our organs is also understood to be dependent on these organs, from the fact that we understand a world as being subjectively conditioned, it does not follow that an objective world is in any case possible. Who prevents us from thinking that subjectivity is real, essential? The 'in itself' is a contradictory conception, to be sure: a 'quality in itself' is nonsense: we have the concept 'being,' 'thing,' always only as a relation concept.... The bad part of it is that along with the old antonyms 'apparent' and 'real' the correlative judgments of value have been propagated: 'of little worth' and 'of absolute worth.'..."[24]

Elsewhere Nietzsche says: "That things have a quality in themselves, quite apart from any interpretation and subjectivity, is an idle hypothesis: it would presuppose that to interpret and to be a subject are not essential, that a thing detached from all relations is still a thing."[25]

Nietzsche's most significant expression of the positivistic world conception is probably given in the aphorism, called "On the Psychology of Metaphysics," where he attack with cutting sharpness the employment of very frequently misused concepts: "This world is apparent: *consequently* there exists a true world; – this world is conditional: *consequently* there exists an unconditional world; – this world is full of contradictions: *consequently* there exists a world that is free from contradictions; – this world is changing: *consequently* there exists a permanent world; – all false conclusions: (blind faith in the reasoning: if there is *A*, there must also be its antithetical concept *B*)."[26]

It is not to be denied that the philosophy of enlightenment possesses a tragic feature. It destroys the old systems of concepts, but while it is constructing a new system, it is also already laying the foundations for new misuse. For there is no theory without auxiliary concepts, and every such concept is necessarily misused in the course of time. The progress of science takes place in eternal circles. The creative forces must of necessity create perishable buds. They are destroyed in the human consciousness by forces which are themselves marked for destruction. And yet, it is this restless spirit of enlightenment that keeps science from petrifying into a new scholasticism. If physics is to become a church, Mach cries out, I would rather not be called a physicist. And with a paradoxical turn, Nietzsche comes out in defense of the cause of enlightenment against the self-satisfied possessor of an enduring truth: "The assertion that the *truth* is *here*, and that an end has been made of ignorance and error, is one of the greatest seductions that there are. Assuming that one believes it, then the will to test, investigate, predict, experiment, is crippled: the latter can itself become wanton, can doubt the truth. The 'truth' is consequently more ominous than error and ignorance because it binds the forces with which one can work for enlightenment and knowledge."[27]

Of these forces, however, at the turn of the century Mach was one of the mightiest.

REFERENCES

* First published in *Die Naturwissenschaften* **5** (1917), 65 at Berlin; then translated in Philipp Frank, *Between Physics and Philosophy*, Harvard University Press, Cambridge, Mass., 1941, and later in the expanded volume of Frank's essays, *Modern Science and Its Philosophy*. – Ed. Reprinted by permission of the publishers from Philipp Frank, *Modern Science and Its Philosophy*, Cambridge, Mass.: Harvard University Press, Copyright 1941, 1949, 1969 by the President and Fellows of Harvard College.

[1] M. Planck, *Die Einheit des physikalischen Weltbildes*, Leipzig 1909.
[2] E. Study, *Die realistische Weltansicht und die Lehre vom Raume*, Brunswick 1914.
[3] E. Mach, *Die Geschichte und die Wurzel des Satzes von der Erhaltung der Arbeit*

written in 1871. An English translation, *History and Root of the Principle of the Conservation of Energy*, was published in 1911.

[4] Planck, *op. cit.*, p. 36.

[5] Published in *Cambridge Philosophical Society Transactions*, 1864, and in *The Scientific Papers of James Clerk Maxwell*, (ed. by W. D. Niven), The University Press, Cambridge, 1890, vol. 1.

[6] P. Duhem, *L'Evolution de la mécanique*, Paris 1903, translated into German by Ph. Frank and E. Stiasny as *Die Wandlungen der Mechanik*, Leipzig 1912.

[7] *Op. cit.*, p. 35.

[8] Study, *op. cit.*, pp. 36, 41.

[9] Quoted by Study, p. 37.

[10] *Ibid.*, p. 37.

[11] M. Planck, 'Zur Machschen Theorie der physikalischen Erkenntnis', *Vierteljahrsschrift für wissenschaftliche Philosophie und Soziologie* (Leipzig), **34** (1911), 497.

[12] Study, *op. cit.*, p. 39.

[13] *Einheit des physikalischen Weltbildes*, p. 34.

[14] *Phys. Z.* **17** (1916).

[15] H. Poincaré, *La Valeur de la science*, Paris, 1905. *The Value of Science*, a translation by G. B. Halsted, was published in New York in 1907.

[16] Abel Rey, *La Théorie de la physique chez les physiciens contemporains*, Paris 1907.

[17] *The Analysis of Sensations and the Relation of the Physical to the Psychical*, translated by C. M. Williams and Sydney Waterlow, Chicago, 1914.

[18] A. Stöhr, *Philosophie der unbelebten Materie*, Leipzig 1907, pp. 16ff.

[19] G. Jaumann, 'Geschlossenes System physikalischer und chemischer Differentialgesetze', *Sitz. Akad. Wiss. Wien*, math.-scientific class. section IIa (1911), and many other papers in the same journal.

[20] *Die Mechanik in ihrer Entwicklung, historisch-kritisch dargestellt*, Leipzig 1883: *The Science of Mechanics: a Critical and Historical Exposition of its Principles*, translated from the second German edition by Thomas J. McCormack, Chicago 1893.

[21] Friedrich Nietzsche, *The Will to Power*, No. 248.

[22] Study, *op. cit.*, p. 24.

[23] H. Kleinpeter, *Der Phenomenalismus*, Leipzig 1913.

[24] Nietzsche, *The Will to Power*, No. 289.

[25] *Op. cit.*, No. 291.

[26] *Op. cit.*, No. 287.

[27] *Op. cit.*, No. 252.

APPENDIX B

PHILIPP FRANK

ERNST MACH AND THE UNITY OF SCIENCE*

In the year 1882 the famous American philosopher and psychologist William James made a tour through Europe and everywhere met scientists interested in his field of work. At the end of October, James came to Prague and met Ernst Mach. James described his impressions of this meeting in a letter he wrote to his wife in America: "Mach came to my hotel and I spent four hours walking and supping with him at his club, an unforgettable conversation. I don't think any one ever gave me so strong an impression of pure intellectual genius. He apparently has read everything and thought about everything, and has an absolute simplicity of manner and a winningness of smile, when his face lights up, that are charming."

I do not want to speak here about the wide activity of Mach in physics, physiology, psychology, history and logic of science, to do which would necessitate a series of papers. Instead, I shall speak about Mach's activity only in so far as he may be considered as one of the spiritual ancestors of the Unity of Science Movement and particularly as the real master of the Vienna Circle.

A hundred years after the birth of Ernst Mach we must not forget or underrate the fact that he is still very much alive today. Surveying the opinions of the scientific workers of today, we find many of them who decidedly reject his doctrine. On the other hand there are a great many scientists, who enthusiastically express their full agreement with him. But among the scientists who are acquainted with Mach's doctrine there are very few who behave neutrally or indifferently toward him. In spite of this fact, opinions about what are the most characteristic features of his doctrine are very different and sometimes even contradictory.

On the one hand Mach is described as the most radical opponent of every attempt to introduce into science factors which have any tinge of a spiritual tendency. He saw even in a concept which for ages has been in physics as usual as the concept of force a detrimental remainder of the obsolete world picture of primitive man, which was animistic and fetishistic. On the other hand we are told that according to Mach our world consists entirely of perceptions or complexes of perceptions; that there is no such thing as matter for the building of a world. For this reason Mach has been proclaimed as a champion of the idealistic philosophy within modern science and as a leader in the struggle against materialism.

To hint at another difference of opinion: on the one hand Mach is said to claim that science has nothing to do but to register observable facts and to sum up these facts by economical and suitable formulas. The scientist has, according to Mach, to be on his guard against daring generalizations, through which an animistic or metaphysical element may insinuate itself into science. On the other hand, the physicists working in their research laboratory accuse Mach of not recognizing the existence of objective facts. According to these people, Mach maintains that there are only subjective opinions of single physicists, but no real facts; there is no real physical world, the exploring of which was supposed to be the goal of the research work of the physicists. For this reason Mach's doctrine is even alleged to have a paralyzing influence on research work and hence on the progress of science. For only the belief in a real, an objective, world can give to the physicist the mental activity and force needed for his difficult achievements.

Whence these different judgments, sometimes even contradictory to each other, about the chief lines of Mach's philosophy? Why is the essence of Mach's doctrine described by different authors in such different ways? The chief reason for these differences is, I think, that philosophers, and sometimes scientists too, endeavor to discuss Mach's doctrine in the language of traditional philosophy. In this language such terms occur as 'idealism', 'spiritualism', 'materialism', 'real objective world', 'subjective opinion of the real world', etc. But the point is that it is impossible to describe Mach's doctrine in this language, impossible to describe it at all in terms of traditional philosophy. If we want to form an adequate conception of Mach's doctrine, we must never forget that he always declined the title of philosopher. Mach has been praised by a great many philosophers for this modesty. But I do not think that it was exactly modesty. He wanted rather to draw a conspicuous boundary line between his own doctrine and the doctrine of traditional philosophy.

In order to understand the chief line of Mach's thought we have, above all, to pay attention, it seems to me, to a passage in the introduction of his book *The Analysis of Sensations.*[1] He explains there what his chief aim was in writing the papers which are commonly described as philosophical papers. Mach starts from the fact that scientists are accustomed, in their special field of science, to make use of a certain system of concepts, or, to put it more exactly, of certain technical terms, a certain technical language which is very suitable within this special field, e.g., within physics. But this special technical language may become very unsuitable and even misleading if it is applied to the description and formulation of the frontier problems which arise when we pass from one special science to a neighboring science, e.g., from physics to biology or to psychology. Mach's own words are: "I do not claim the title of a philosopher. I want only to take, in physics, a standpoint which does not have to be abandoned immediately when we look over into the field of another science. For all the sciences ultimately form a whole. What I am stating, I am perhaps not the first to state. Furthermore, I do not want to bring forward my explanation as an extraordinary achievement. I think, rather, that the same line would be taken by anyone who tried to survey a field of science that is not too narrow." According to Mach this desire to make use of a unified mode of expression in all fields of science is a consequence of the economical design of science. This design implies the comprehension of as many facts as possible by the simplest possible system of propositions.

Since Mach dealt in his papers with so many different kinds of problems in the fields of physics, physiology, and psychology, a great many scientists did not discover what the chief trend of these papers was. In order to find the clue to them, we have to read attentively another passage in the same introduction. Mach says there expressly: "The basis of all my investigations into the logical foundation of physics as well as into the physiology of perceptions has been one and the same opinion: that all metaphysical propositions must be eliminated, because they are idle and disturbing to the economical design of science." And Mach's famous book *Mechanics and its Development*[2] begins with the sentence: "The tendency of this book is an elucidatory or, to put it more accurately, an anti-metaphysical one."

In the reports which advocates of traditional philosophy have given on Mach's teaching, you may often read that Mach's chief doctrine was to maintain that the world consisted of perceptions and not of material particles. But you see from the passages which I have just quoted, in which

Mach expressly describes the chief aim of his investigations, that these investigations were quite unconcerned with such problems as whether the world consists of perceptions or of matter. This is rather the typical way in which traditional philosophy likes to put a problem. It is just this way of putting a problem that Mach emphatically rejected.

Hence Mach's chief tendency may be described by the slogans 'unification (i.e., economical presentation) of science' and 'elimination of metaphysics'. We shall find that these two aims are very closely connected with each other. And we shall see that the most widely popularized doctrine of Mach, according to which the real world consists of perceptions, was never formulated by him in this metaphysical way. If we try to find out what he really thought when he maintained that our world is built up of perceptions or complexes of perceptions, we shall soon see that this phrase 'built up of perceptions' is in Mach's eyes not in any sense a statement concerning a property of the real world, but only a useful means for the unification of science and the elimination of metaphysics. It would be a misunderstanding of Mach's aims to believe that the construction of the world by perceptions, which was merely a means to an end, was the true end of his philosophy. Many of his philosophical interpreters stick fast at this means to an end, 'the perception language', and neglect the real purposes of Mach's doctrine, the unification of science and the elimination of metaphysics. For it is just by the presentation of some scientific results in terms of metaphysics that the unification of science is, according to Mach, gravely imperiled and sometimes even frustrated.

If one describes physics as the science of matter, biology as the science of life, psychology as the science of the mind, sociology as the science of the collective mind, metaphysical concepts or words such as *matter*, *life*, *soul*, *collective soul*, etc., are introduced, and it is obvious that words like *matter* and *soul*, for example, are probably not reducible to the same terms. It is easy to prove that the introduction of expressions of this kind renders impossible the representation of our experiences by a unitary system of terms, in other words, renders the unification of science impossible.

To remove these difficulties Mach suggested the formulation of the laws of physics as functional connections between perceptions such as *green*, *warm*, *hard*, etc., including also, of course, the space and time perceptions. Every physical experiment consists in observing how the alteration of some perceptions is connected with the alteration of others. If no perceptions touching our own bodies intervene, e.g., no altering of perceptions by intoxication of our nerves, we are in the field of physics. If we observe the connections between perceptions, including perceptions arising from alteration in our own bodies, we are in the field of physiology or psychology. But you see that we can no longer prove the impossibility of a unified language of science if we start from Mach's perception language instead of the metaphysical terminology of traditional philosophy, and, we must admit, sometimes even of traditional science.

On the contrary, if we start from Mach's standpoint and formulate all scientific propositions in terms of perceptions, the unification of science becomes possible. Mach never maintained that our world consisted of complexes of perceptions, but that every scientific proposition was a statement about complexes of perceptions. Whether it be a proposition of physics, biology, or psychology, it can only be proved or refuted by comparison with observation. But we can formulate about observations only such propositions as contain perception terms as predicates: e.g. *green*, *warm*, etc., *joyful*, *painful*, etc.

We see that, according to Mach, the unification of science is possible, but only by formulating all scientific propositions as propositions about complexes of perceptions, in the widest sense of this word. Every proposition that states something about our observations contains terms like *green*, *warm*, etc., as predicates – perception terms, as Carnap calls them. A proposition that is not reducible to propositions containing only perception terms as predicates cannot be checked by experience; it is a metaphysical proposition. Hence to Mach the expression 'elimination of metaphysics' means the elimination of all sentences that are not reducible to sentences containing only perception terms as predicates. The elimination of metaphysical propositions from science leaves only sentences of a homogeneous type; namely, sentences with perception terms as predicates. Therefore, if we demand of science an economical representation of our experiences, i.e., a representation by a unified system of concepts, we must admit only propositions that are reducible to propositions containing only perception terms as predicates.

This is the real meaning of Mach's doctrine that all propositions of science deal with perceptions. He did not want to make a statement about the question of what the world consists of, but he wanted to point out how the propositions of science had to be formed in order to make possible a unification of science. His result is this: The unification of science is not possible except by the elimination of metaphysical propositions. Then only propositions of a homogeneous type remain. Hence we can form from them a coherent logical system.

The elimination of metaphysics from science was not for Mach, as we now understand, a demand arising from some anti-metaphysical mood, but the only means of making possible the unification of science. According to Mach, metaphysics must be eliminated "because it is contradictory to the economical function of science".

A lot of people were puzzled because Mach's philosophy, which was supposed to be a sort of idealism (similar to the philosophy of Bishop Berkeley), changed so easily, or – to speak in terms of idealistic philosophy – degenerated so easily, into physicalism. We have seen that even the Vienna Circle went over very quickly from the phenomenal language, used, following Mach, by Carnap as well as by Schlick, to the physical language claimed by Neurath. In physicalism, which now plays a great role in all papers starting from the viewpoint of logical empiricism (in the most coherent and exact way in the papers of Carnap), a language is used that seems very near to materialism.

So it has been for a great many philosophers a riddle and almost a source of irritation that the opinions within a group claiming to possess a particular sense of logical consistency could oscillate so easily between the extreme poles of human thinking, idealism and materialism.

But this antithesis that according to traditional philosophy exists between materialism and idealism is not, according to Mach, a scientific antithesis. Mach disliked making use of terms such as 'idealism' and 'materialism' and if he did use them it was only to reject them. Though he rejected materialism as well as idealism, this rejection does not mean that he tried to take a mediating standpoint between idealism and materialism. For him, both idealism and materialism are systems of metaphysical propositions, not scientific theories. For they can be neither proved nor refuted by experience. He had what one might call the instinctive aversion of a genuine scientist to the use of vague terms like 'idealism' or 'materialism' in science. This aversion sometimes induced him to make statements against one or the other metaphysical system.

And his statements were often misinterpreted as statements on behalf of the other system, just as if, by rejecting one sort of metaphysics, he advocated the contrary metaphysics. From Mach's point of view the question of 'idealism' or 'materialism' cannot be put as a real scientific problem. Every attempt to exploit the achievements of science to bolster up idealistic or materialistic metaphysics is from the beginning doomed to failure.

What Mach felt instinctively we can today formulate in words, if we take the standpoint of logical empiricism, as it was formulated very precisely in Carnap's *Logical Syntax of Language*[3] and in his paper 'Testability and Meaning.'[4]

The transition from the supposed idealistic to the physicalistic conception of science took place so smoothly within the Vienna Circle because, according to the doctrine of logical empiricism, the question to be put was not whether idealism and materialism were right opinions about the real world, but only what language was the most suitable for giving an economical and unitary account of our experiences, whether the phenomenal language or the physical language. Since either of these languages may be more suitable within a limited field than the other, the choice of an accepted language has nothing to do with the question whether our real world consists of perceptions or of matter. What is of essential importance is only the question whether we believe that it is possible to comprehend all fields of science in one and the same language. If the unification of science, in this sense, is held to be possible, as Carnap and the adherents of the Unity of Science Movement maintain, it is of secondary importance whether this unification be achieved in terms of perceptions, as Mach believed and Carnap proved to be right in his first paper 'The Logical Construction of the World',[5] or whether the physical language is to be introduced, as Carnap proclaims in his recent papers in accordance with Neurath's suggestions.

The essential alternative for our conception of science is rather this: Do we, in accordance with traditional philosophy, maintain that the question whether the real world consists of matter or of perceptions and other questions like it are scientific questions, or do we, with Mach, eliminate metaphysical questions of this sort from science as disturbing its economical character and put the question just mentioned in the way it is put by logical empiricism?

Then we ask, what language is the most suitable as the language of a unified science. From this point of view, the metaphysical question seems to be, as Mach expressed it, an idle one. And the question as logical empiricism puts it, whether the phenomenal language or the physical language is more suitable as the language of unified science, ceases to be a question of profound metaphysical importance and becomes a question of convenience. It is perhaps comparable to the question of what system of symbols is the most suitable for the introduction of a unified symbolism into logic.

If we want to describe Mach's role in the history of human thought and in the development of science in a particularly comprehensive and impressive way, we can do it, it seems to me, by a clear-cut antithesis. The traditional conception of science is connected with a certain opinion about the importance of metaphysics to science. According to this conception, there are two methods of science:

(1) The first method is restricted to registering facts, including empirical rules summarizing facts. The adherents of this method of scientific activity take care not to introduce sweeping generalizations and hypotheses, because by them a metaphysical element may easily creep into science. This kind of scientific activity possesses, according to its adherents, the advantage that all propositions admitted by science are secured by experience or logic and are all clear and intuitive. The scientists of this group are anxious not to introduce vague phrases. But by proceeding in this cautious way only special sciences of small extent are created. Physics and biology and psychology have each collected a lot of facts and rules, but there is no link between these special departments. This concept of science is often referred to as the 'positivist' conception of science. This conception is not at all in accordance with the conception of the so-called 'logical positivism', neither does it agree with Mach's doctrine. This pseudo-positivist conception of science does not satisfy our aspirations toward the unity of science.

(2) Therefore, beside this positivist conception, or more exactly, over it, the metaphysical conception of scientific activity has always existed, because it has been supposed to be better adapted to satisfying our desire for a synthesis of knowledge and for the unity of science. According to this conception we can reach the unity we are striving after by the introduction of daring metaphysical generalizations and hypotheses. By

means of these metaphysical generalizations the separate sciences can be summarized and unified into a unitary science. The general principles of this unified science are, of course, metaphysical propositions. The most famous metaphysical system, which was supposed to comprehend and represent all special sciences, was the system of Hegel. In this system all separate sciences, such as mathematics, physics, biology, are presented as steps in the self-evolution of the absolute spirit. If you want to have an example of the sort of metaphysical proposition that serves to achieve the unification of science, you can refer to Hegel's fundamental theorems of dialectics, e.g., "every quantity if sufficiently increased, turns into a quality". This theorem is supposed to be valid in physics as well as in biology, in biology as well as in history. This theorem is perhaps particularly interesting, because it continues to play a great role and is not confined to the adherents of Hegel's idealistic metaphysics. This theorem and many like it are taken over into dialectical materialism, the official philosophy of the Soviet Union of today. By means of propositions of this kind the barriers between the separate sciences are broken down, the unification of science is achieved, but only at the price of introducing large systems of very vague propositions. With regard to such metaphysical propositions a general agreement among scientists would never be reached.

The opinion of traditional philosophy and traditional science has been for ages that scientific activity has only these alternatives: either that only theorems which can be proved by experience or logic are recognized as legitimate in science, in which case the separate sciences remain separated from each other by insurmountable barriers, or that we admit the introduction of metaphysical propositions, in which case the unification of science can be achieved, but we have to deal with propositions that will never be recognized by all scientists.

Or to put it still more concisely: either *the renunciation of the unification of science* or *the introduction of metaphysical propositions into science.*

The great importance of Mach's activity lies in the fact that he declined to recognize these alternatives. He proclaimed, rather, *the unification of science by means of the elimination of metaphysics.*

It is just this sentence that is the clue to the understanding of Mach's doctrine, of his papers, which seem to deal with so many subjects and such different fields of science. What always mattered for Mach was the opportunity to accomplish the program that we have just outlined. And it is just this program of Mach that we may adopt as the program of our 'Unity of Science Movement', of our Congresses and of our Encyclopedia. If Mach's centenary has been celebrated by so many physicists, physiologists, psychologists, and historians of science we may take a special pride in these celebrations, in which we have a special right to honor him as one of the spiritual ancestors of the 'Unity of Science Movement'. For, I think, within our movement, the harvest of the seed scattered by Mach is particularly rich and is in the strictest accordance with his true intention.

REFERENCES

* First published in *Erkenntnis* 7 (1938), 247 at The Hague; then translated in Philipp Frank, *Between Physics and Philosophy*, Harvard University Press, Cambridge, Mass., 1941, and later in the expanded volume of Frank's essays, *Modern Science and Its Philosophy* – Ed. Reprinted by permission of the publishers from Philipp Frank, *Modern Science and Its Philosophy*, Cambridge, Mass.: Harvard University Press,

[1] *Beiträge zur Analyse der Empfindungen*, G. Fischer, Jena, 1886; English translation, Chicago, 1914.

[2] *Die Mechanik in ihrer Entwicklung*, F. Brockhaus, Leipzig, 1883; English translation, Chicago, 1893.

[3] *The Logical Syntax of Language*, Harcourt, Brace and Co., New York, 1937.

[4] *Philosophy of Science*, 1936 and 1937.

[5] *Der logische Aufbau der Welt*, Berlin 1928.

RICHARD VON MISES

ERNST MACH AND THE EMPIRICIST CONCEPTION OF SCIENCE

*An Essay for the Centenary of his Birth, February 18, 1938**

A person endowed with normal senses, pursuing a normal middle-class occupation, *sees* before his eyes the surface of the earth as an infinitely wide surface, only occasionally interrupted by rising ground and hollows. He *feels* the earth as being firmly grounded and at rest, and he *knows* that the sun is rising in the morning in the celestial vault and setting in the evening. He has some vague notion dating back to his schooldays, something rather unclear, never completely comprehended, that all this is only deception. The earth is round and if you dig under your feet and go on digging always in a straight line you finally end up at the feet of another person who carries his head downward without having the slightest feeling of being in a precarious position. Also, the world is not at rest, but turns around its axis without let-up, and does so with such vehemence that, thanks to this rotation, we travel almost thirty kilometers per minute, taking with us all that is around us, houses, towns, rivers, and forests. And as if this were not enough, the whole globe sweeps through the universe at a speed sixty times faster than that. Only the moon – oh, well, everything is different with the moon: it really circles around the earth, – here is truth, while with the sun all is only illusion, but, oh! – such a convincingly strong illusion.

A person who thinks normally does not doubt what he learned properly between the sixth and tenth year of his life, he accepts it, and in time, consciously or unconsciously, it become a basic part of his view of life. He believes consequently that there are two kinds of worlds, one which we perceive with our senses, but which is more or less illusion and deception, and another one, behind it, which is the true, the real world, which is disclosed by the perspicacity of the scientist. There are various ways of coping with this pair of opposites. In moments when we would like to lean on what we immediately observe, we look on the world of sense perception as reality, as *praxis* and, somewhat contemptuously, we call the true world 'theory'. But, at other times, when we happen to be annoyed with someone who does not follow the flight of our thought, we brand his point of view as vulgar empiricism or even as positivism, and we hold up to him the pure world of ideas, or 'idealism'. No one who reflects on his experience will find it easy to come to terms with this dichotomy. At the age of seventy-one, ten years after he finished his treatise on colors, Goethe, still restlessly searching, wrote under the title *Bedenken und Ergebung* (*Doubt and Surrender*)[1]: "Here we encounter the real difficulty of which we are not always clearly conscious, namely, that between idea and experience there appears to exist a certain gap which we try to bridge with all our might, but in vain. Nevertheless, we continue our never-ending endeavors to overcome this hiatus by means of reason, intellect, imagination, faith, feeling, and delusion, and if all this does not succeed, we resort to nonsense." It was Goethe's fate to recognize the problem, to circle it unceasingly, to surround it with brilliant aphorisms, but never really to elucidate it. His whole, infinitely painstaking scientific activity suffered from this uncertainty, of which he was aware.

Others, of a more robust character, with talents restricted to the natural sciences, were able to overcome these difficulties or else to ignore them. This accounts for the great achievements in the natural sciences of the 17th and 18th centuries. Newton is a unique phenomenon in this respect. With unerring certainty he created the basic outlines for the construction of a determinist physics, a base which lasted for centuries. At the same time, he wrote a considerable number of essays on legal, historical, and, above all, theological topics; for him there was no such thing as an epistemological problem. In his main work *Philosophiae Naturalis Principia Mathematica* (*Mathematical Principles of Natural Philosophy*), he devotes a bare three pages to the *regulae philosophandi*, the rules of reasoning in philosophy. They are found at the beginning of the third book which contains a description of the stellar system. The hasty style of these pages shows how much the author hurried to get on with the discussion of the *phaenomenon primum*. In terms of substance, the four rules do not offer much more than what is expressed in the often cited slogan: *hypotheses non fingo*. Newton was as little able to adhere to this rule as anyone who establishes a theory ever is or ever will be. But he did have an absolutely sure scientific instinct which enabled him to distinguish between empty speculations and testable hypotheses. It never occurred to him to reject the latter, and his careless formulations only testify to the fact that he was not burdened with epistemological problems.

"To introduce the experimental method in the intellectual realm", is how David Hume, fifty years after Newton, formulated the problem in his first work, which was later shortened, rewritten, and adapted to the tastes of the public, a work widely known since under the title *Enquiry concerning Human Understanding*. What really interested Hume was to establish lucidity of thought in *all* fields in which the human intellect was active, and he succeeded in this endeavor to a degree which was admirable for his time. In this respect, Newton could serve as a model in his own field. In other matters, especially where ethical and religious problems were concerned, the deck had to be cleared of an immense mixture of confused notions, of fixed manners of speech, and of misguided judgments. Hume's theory of knowledge contained elements which might have cleared up difficulties which kept bothering Goethe during his whole life. It seems, though, that the poet and scientist never had any contact with the cool sceptic (who seemed so alien to Goethe), who perhaps just because of the fundamental difference between them might have had beneficial complementary value for him.

The path Hume took could have led straight forward, had he not turned once more to climb the kind of peak that grants a lovely view to the wanderer and invites him to linger a while, but from which he nevertheless has to turn back if he does not want to fall down the precipice. Kant, the great enlightener who admitted that Hume had interrupted his dogmatic slumber, was called the 'Prussian Hume' by the mystic Hamann. Once aroused, Kant directed all his efforts towards replacing dogma by the most strictly structured system. Little was left of the human breadth of Hume's work, of his genius in understanding all that is humane and all that concerns the intellect. The *Prolegomena to any Future Metaphysics* show more clearly even than the *Critique of Pure Reason* what, in the last analysis Kant's system aimed at: to prove – in a grand style and with a display of highest scholarship and extreme perspicacity – that it is after all the schoolmaster who is right. True, all hitherto prevailing metaphysics is to be rejected as nonsensical and sterile; true, truth can be derived only from experience, but, in regard to what we perceive, what we know, there is and remains only appearance, semblance dulled by the imperfection of our senses, behind which in fathomless distance, forever unknown to us, there exists, free from all impurities, the pure Idea. This skillfully constructed system of thought which makes abundant use of the most common artistic means, that of symmetry, allows various, even contradictory, interpretations.[2] It teaches us that, on the one hand, it is quite impossible to know "prior to all acquaintance with things, that is to say, prior to their being given to us" what they are like in terms of intuition, but that, on the other hand, pure reason furnishes *a priori*, i.e., prior to all experience, apodictically valid knowledge concerning the forms of intuition. It is on this sharply drawn edge where, for Kant, pure mathematics, pure natural science, and pure metaphysics are possible.

It is well known what effect the Kantian system had on some of his contemporaries and what influence it exerted throughout a full century. The main tenets of transcendental critical idealism gradually came to be looked on as unassailable, in spite of a great variety of interpretations and in spite of the creation of new systems which here and there contradicted the master. In time the combined action of disciples, interpreters, critics and moderate opponents produced a body of doctrine which constituted the "academic philosophy (*Schulphilosophie*) of the nineteenth century". So far, no equally strong and equally well organized opponent has come forward to confront this school. Nietzsche was not sufficiently interested in the theory of knowledge. Comte, who in many respects took on more than he could handle, had little influence outside of France and Western Europe.

We know what pains Goethe took in his old age to assimilate something of Kantian philosophy. He "liked to take that side which does most credit to man" and he thus accepted the theory of synthetic *a priori* judgments. But one has the impression that he is not quite at ease with it, and it is evident that this doctrine was of little help to him in solving his epistemological problems in the natural sciences. There is some doubt whether any great natural scientist was ever in fact a Kantian, but there is none that the education of all of them during the last ten or twelve decades was strongly influenced by Kant's system. At the age of fifteen, Ernst Mach found in the library of his father a copy of Kant's *Prolegomena to any Future Metaphysics*, and he later termed this event "a stroke of special good fortune" (*The Analysis of Sensations*, Dover edition, N.Y., p. 30 [1959]), since "at that time the book made a powerful and ineffaceable impression upon me, the like of which I never afterwards experienced in any of my philosophical reading". Mach went on to explain: "My relation to Kant has been peculiar. His critical idealism was, as I recognize with the greatest gratitude, the starting point of all my critical thought; but it was impossible for me to retain allegiance to it." (*op. cit.*, p. 367)[3]

When Mach finished his studies at the university in 1859, Darwin's *Origin of Species* had just become known. A year later Fechner's *Psychophysics* appeared, and not long before that Helmholtz had placed the general principle of conservation of energy into the center of physical research. The young physicist found himself in a situation which presented him with tasks that were to occupy him all his life. Inspired by Kant's and Herbart's philosophy, particularly by the latter's attempt to create a mathematical psychology, he had developed a fine ear for the difficulties of many classical doctrines presented as self-evident. In various branches of the natural sciences, fundamentally new ideas were arising which claimed validity beyond a special field, even beyond the natural sciences; but there were ideas which could hardly be reconciled with each other, and which in any case could not be subsumed to the traditional methodological and philosophical views. To be sure, the battle against purely speculative philosophy of nature, on the order of Hegel and Schelling, had already been fought at that time, so that fortunately Mach was spared from exhausting his energy in polemics. A sure instinct guided him away from the daily routine of natural science research which merely tended to increase the accumulation of individual unrelated findings and information. Freely and independently he chose that method of work which he found adequate for himself, which however, seemed diametrically opposed to that of the normal natural scientist. The young man recognized that the most reliable means to elucidate methodological questions was to study them historically and throughout his life he adhered to this point of view. To trace the growth of ideas as they developed in the minds of great scholars from the earliest times constituted the observational, experimental part of his research activity, and this was the basis for far-reaching critical work which led to a complete renewal of all views on the nature of knowledge. This is how Mach became the man who is recognized today as *the most powerful and, for our own time, the most typical philosopher of enlightenment in the last generation.*

In 1866, Mach sent the first result of his historical-critical labors to Poggendorf's *Annalen der Physik*, which already was the leading professional journal in physics. In this essay, Mach attempted to establish a new conception of mass which already contained the germ of many ideas that matured later. The unpublished manuscript was returned to him one year later. Whoever looks today through some of the volumes of the *Annalen* for that time, can only marvel at the completely uncritical way in which articles without any relevance, even absurd ones, were selected for publication. Nevertheless, there clearly existed an insurmountable barrier erected by faith in authority, by education along the lines of the orthodox academic philosophy school of the day, and by the naive notion that certain statements such as Newton's principles were no longer under discussion. Later in life, Mach once more encountered such an expression of unmitigated, self-assured naïveté which – we do not wish to deny it here – may have some pragmatic value: one of the greatest contemporary physicists turned against Mach with the argument that whosoever does not believe in the reality of atoms is simply not a physicist and cannot claim scientific validity for his work. In what is perhaps his only polemical statement, Mach said, in his replay of 1910: "We see that the physicists are well on their way to becoming a church; they are already mastering the church's most familiar methods". He declined to join the community of the faithful, he preferred freedom of thought. The evolution of physics since 1910 has sufficiently shaken the dogmatic faith of the physicists' church, current at the beginning of the century, and, some time ago it has become clear to every thinking person how much the *freedom of thought* created by Mach has contributed to progress.

Without entering into problems of physics in the narrow sense of the term, we shall attempt to describe some of the general fundamental ideas by which Ernst Mach contributed decisively to the clarification of what is generally known as the *problem of knowledge* (*Erkenntnisproblem*).

When Kant poses the question: how is pure natural science possible? he does not think of physics or chemistry as historical reality but, in a certain sense, he thinks of natural science as a Platonic Idea. For Mach, science is first of all something given here and now, an actual manifestation of life of the human species. He looks on the history of science from a biological point of view as part of the story of man's evolution. Prior to any scientific research in the narrow sense of the term, man instinctively collected experiences and formed some primitive concepts by creating an everyday language. In full agreement with Mach's ideas, Nietzsche once said: "Originally it is language which builds concepts; later it is science". Both language and science later on evolve in such a manner that man's thoughts and his means of communication adapt themselves gradually to the facts which are known or experienced. Scientific thought continues a biological evolution, which began with the first most primitive expressions of life. There is no sharp division between pre-scientific and consciously scientific thought, nor can there be any question in any field whatsoever of a final conclusion, of a definite form of knowledge. For Kant, the Copernican theory was a 'one-time' revolution which wiped out a former error and established a forever valid absolute truth in this particular domain. Mach assumed, and later developments showed him to be correct, that the Copernican doctrine is likewise only a certain phase in our knowledge of the universe which may be modified again by assimilation of new facts, perhaps even by findings in an entirely different field.

For the task of scientific research is not limited to the adaptation of thoughts to experiences (*Erlebnisse*); but it comprises also the mutual assimilation of ideas to each other. Man wants to comprehend as large a range of phenomena as possible by means of uniform consistent ideas, and by *comprehend* we mean, to be able to retain them in his mind and communicate them to others. This part of research is what is called, in the narrow sense of the term, *theory*. It is the logical purification of thought, which we might also call the real philosophical task proper if we do not mean by philosophy mere academic philosophy (*Schulphilosophie*). Thus, from systematic observations of freely falling bodies, Galileo drew conclusions which Newton later connected with observations of celestial motion. It was the endeavor to find a unified conceptual design fitting the whole range of all phenomena of motion which led to Newton's 'philosophical' principles of mechanics, especially the general idea of a force determining acceleration (not speed), and eventually the specific statement of the law of gravitation. For two centuries, these statements met all demands resulting from the constantly evolving comparison between theory and observation, until finally certain divergencies appeared, which required a change in the theory. For a long time, an effort was made to change details but yet to maintain the most general principles of Newtonian physics. Only when the vast majority of physicists agreed that this was no longer possible did Einstein's theory of relativity gain acceptance, a theory which contains a modification of the classical principles of motion – very difficult to make intelligible to the layman – but which simultaneously fits all known phenomena. We cannot reconcile this development with Kant's conception of science, according to which the specific law of gravitation is a proposition which "cannot be conceived *a priori* but must be proven by inductive methods", while, on the other hand, in order to arrive at this induction, "Newton had to *apply* certain general principles of mechanics: for example, he had to deduce from the principles of the mathematical philosophy of nature the hypothesis that all changes in motion are effected by continuously accelerating forces...." (see Leonard Nelson, *Abhandlungen der Fries'schen Schule*). Mach rejected as unfounded and contrary to the facts a distinction in principle between *propositions* derived from experience and *basic laws* which are apprehended *a priori* and which, endowed with eternal validity, are above mere propositions (*Lehrsätze* vs. *Grundsätze*). All we can say about the world – individual assertions, rules, natural laws, theories, principles – is constantly subject to control and re-examination, and there is always the possibility of a modification through experience.

In order to refute a naive kind of empiricism (which should not be confused with Mach's biological conception of science) people often say: "The statement that there can be no apodictic truth is in itself an apodictic assertion and therefore self contradictory." But this statement was not one that Mach ever made. For him science is a part of nature, one of the manifestations of life which characterize the human species, and the kind of statements he makes in his theory of knowledge about the particular nature of this phenomenon of life is of exactly the same type as those made in the natural sciences about other phenomena. Among the propositions with which we try to describe our impressions – in the widest sense of the word – we cannot single out one special class of propositions as derived from an essentially different source and therefore entitled to a substantially longer duration of validity than others. All our previous experience teaches us that. Nothing can be done either with 'immediate insight', or with the notion that pure intuition or similar sources of information provide the evidence of principles. For we know that almost every new insight – later looked on as a firmly established principle – appeared, when it was first introduced, as strange and contrary to good sense. It took generations to get used to these insights and to acquire for them *through habit* the place of evident, self-understood truth. But now we have come to a somewhat different question which must be more closely examined.

The problems of knowledge, as they are reflected in the works of philosophers since the days of the Greeks, have in the end a single definite source: the existence of mathematical truth which, according to all appearances, cannot be doubted and is absolutely certain and exact. If we were to be slightly disrespectful, we might say that the existence of mathematical propositions had turned the philosophers' heads. Over and over again, in slightly different ways, sometimes clearly stated, often only silently assumed, academic philosophy sets forth trains of thought as follows: mathematics succeeded in reaching final apodictic truths which no reasonable human being can doubt; therefore, mathematics is the model for all sciences, and they must emulate it in order to obtain for their own propositions the same, or in any case a similar certainty and exactness. There must be, at the very least, general philosophical principles which can be proved *more geometrico demonstrata* and which thereby become unassailable. People often cite Kant's remarkable statement that the amount of genuine science in any particular natural science is equal to the amount of mathematics contained in it. Let us not argue that Kant was a special adherent of the mathematical method, which did not lack detractors. Schopenhauer, for instance, who was certainly one of the most vehement philosophical opponents of mathematics, merely finds the "manner in which it proceeded strange and in fact wrong". He wants to shift the emphasis from mathematical proofs to pure intuition as the real source of mathematical knowledge, but he too does not doubt its infallibility; and he refers to it as the measure of the reliability of philosophical propositions, for example, of the one which he valued most highly, the principle of sufficient reason. The endeavor to establish a clear link between the so-called 'necessity' of the relation between cause and effect and the necessity of geometrical theorems is even more marked in Schopenhauer's work than in that of Kant.

Something Goethe is reported to have said in the year 1826 suggests more clearly and more correctly than the statements of most professional philosophers what the notion of the infallibility of mathematics really is about. "Mathematics", he said, "is quite erroneously reputed to furnish infallible conclusions. All its certainty is nothing else than identity. Two times two is not four, it is just twice two, and we call it for short four. But four is absolutely nothing new. And so it goes with all mathematical conclusions, only we lose sight of this identity in higher formulas".[4] Here, we might add a qualifying note and say that four is not only twice two but also, for instance seventeen less thirteen, or one sixth of twenty-four. What is really difficult to explain is the identity of 'four' in all these cases. Goethe, nevertheless, reveals himself here as a precursor of Mach's positivism when he shows with decisive and enlightening insight that mathematics, *inasmuch as it furnishes infallible conclusions*, says nothing about actual facts, about something outside itself. Mathematics, therefore, does not furnish what Kant means by *synthetic* judgments. Einstein makes this even more clear and expresses himself in a manner which is closer to Mach's point of view when he said, in 1921, in his address of acceptance of the Nobel prize: "As far as the laws of mathematics refer to reality, they are not certain; and as far as they are certain, they do not refer to reality."[5]

If we want to go beyond familiar inadequacies, we must give up, first of all, the entire notion that arithmetic plus geometry equals mathematics. This conception has its origin in school instruction and can only be justified historically and didactically. Geometry, the theory of spatial phenomena, is, as Mach insisted over and over again, part of the natural sciences, just as mechanics or the theory of heat or the theory of electricity. Without experience, without observation, we can make no accurate correct statement about spatial forms, just as little as about the motions of celestial bodies or the thermal effects of an electrical current. Only rather simple observations, however, are needed to construct a theory of space, and these observations were made quite early, probably, to some extent in very early times of human history. These circumstances make it appear – and misled people to make the assumption – that the results of these observations are ours, independent of all experience, that they are imprinted on our mind, that, in other words, they constitute a *necessary way of thinking*. But we know today how to construct, without contradictions, a geometry which not only does not satisfy all postulates of Euclid but even partially contradicts them. Only by means of continued observation – which up to now has not been terminated – can we decide what kind of geometry is best suited to represent the actual relationships in a larger cosmic space. Thus the infallibility and certainty of geometry is not worth so very much.

On the other hand, in geometry, just as in mechanics or other branches of physics, mathematical inferences are used to pass from initial assumptions to theorems or to obtain certain particular results. The simplest of such inferences are familiar to everyone, they are used constantly in daily life, and only if they are accumulated, linked, and concentrated, do we speak of a difficult mathematical deduction. We do not want to go into too much detail, and we shall therefore admit that, within certain limits, these methods of inferences are reliable, or rather, they are generally accepted. They permit us to pass from an accepted initial proposition or a premise to a specific conclusion, but no matter how perfect a method of inference is used, it cannot lead to any statement which would be independent of what has been put into the initial proposition. If, therefore, we apply mathematics in the field of geometry, or in some part of physics or in any other field of knowledge, or in an everyday situation, we can never obtain results which do not contain *at least* all the uncertain and earthbound elements peculiar to the major premise taken from experience. We cannot use mathematics for anything else than to lead us from one proposition – which expresses results of observation and does not belong to mathematics – to an assertion which is likewise outside mathematics and refers to something that can be experienced. Spinoza's ethics for instance *more geometrico demonstrata* would not be free from the uncertainty contained in the initial propositions of the deduction, even if the method of inference used were mathematical (which it certainly is not).

We should not underestimate the importance of these new insights into the relation between geometry and experience. If there is no longer a superstitious faith in the apodictic character of geometry, innumerable illusory problems disappear completely, problems which have been lying like a heavy weight on the mind of man struggling to gain knowledge. What disappears first is the fundamental question of critical philosophy: how are synthetic judgments *a priori* possible? how can we explain the 'necessity' of causal relationships? – and similar questions. For those whose mind has finally been freed from the ghost of infallible, absolutely valid geometry, there can be no longer any doubt that the link between cause and effect is known only *through experience*, that this relation can never be formulated except as an experienced fact. But if we cannot render particular concrete causal relationships in apodictic terms, how much less is this possible when we make such a general assertion as, e.g., that the description of nature *must* take a causal form. The newest developments in physics suggest fairly clearly that more general forms of description have to be admitted, thereby demonstrating a freedom of thought for which there had been no room within the framework of academic philosophy. A great and longwinded process of liberation and humanization is taking place; it started with David Hume, or reached a first summit with him, and it was continued most successfully by Mach.

No one who has kept a sound historical instinct is going to deny the imposing perspicacity, great constructive systematizing force and intellectual attitude which found expression in some works of classical philosophy. But we tend to believe that these older endeavors to elucidate the problem of knowledge have gradually, in relation to us, taken on for us the role of alchemy and astrology, the precursors of modern science which, in spite of a fundamentally wrong approach, produced some useful results of value for later scientific research. But all our sympathy for the aesthetic values of the intellectual past cannot prevent us from ranging ourselves on the side of chemistry and astronomy, and to keep our distance from *today's* alchemists and astrologists. This is what Mach taught us.

We are approaching again our point of departure, the dilemma of the world of appearances of our senses and the 'true' world of science. Is 'red' a sensation of color or is it in reality a vibration of the ether of so-and-so many trillion frequencies per second? What is left of any object or body in our environment if we withdraw from it all qualities subject to change such as color, temperature, extension etc.? An unchanged nucleus? A 'thing in itself' (*Ding an sich*)? Is the human individual anything more than the sum total of all that we perceive, his appearance, his character, his actual conduct in all of life's situations? Mach's analysis of knowledge provides an approach to all these questions; its effect is liberating because it clears away accumulated debris and it provides release by opening our eyes to an unprejudiced view.

The fact that we call things by the same name even while they change misleads us to assume that something remains even if we abstract from all observable qualities. In very simple cases we are familiar with the nonsense of this assumption. If we have a knife and we change the blade and later the handle and we continue to call it 'the same knife' we know that this is not for a very profound reason; it is merely habit, a comfortable way of expressing ourselves. It is somewhat different when, as Goethe recalled, in the house of his father things keep changing: if successively all the walls and staircases and floors are renewed, if the interior and the exterior of the house are completely changed, and if, at the same time, the family which continues to live in it and all other inhabitants of the town speak of *the* house, then we have somehow the feeling that there remains something like an essence of a higher kind. It is a fact that such confused feelings, emotional needs so to speak, have given rise to the most diverse philosophical theories which culminated in Kant's concept of the *thing-in-itself*. If we approach the problem in a natural manner and do not get ourselves tied up in a scholastic knot, we find that if we want to give a definition of '*this*' house, we must take into account not only data on materials, building plan, measurements and appearance of rooms, in short all that belongs to its material aspect, but we also have to include what can be said about its history, its origin, its use, the practical activity and the objective and psychological attitude of its inhabitants and of all others who use it. All these data form a unit, an identity; in principle, they are all of the same kind, namely, derived from observation and verifiable by observation. On the other hand, the sum total of these data is *all* that constitutes the concept '*this*' house'; there is not in addition or behind it the *house-in-itself*, a mysterious and forever inaccessible something which is preserved even if *all* determining parts which we mention here change and are replaced by others. If we change only one kind of the parts, for instance the physical ones, we are free to continue to speak of the same house. Only the practical needs of communication, habit, and tradition determine the limits within which we maintain a verbal designation in spite of changes in a part of the aspects which characterize an object. Of course, we also can make this linguistic attitude the subject of a study based on observation – but the stability of the word, quite certainly, does not connote an unchangeable 'idea' realized only imperfectly in the world of appearances. *The house is what we perceive it to be and what we observe in connection with it.*

Sense impressions are not deceptive and they do not deceive us: only through sense impressions is the world given to us. The fact is though that sense impressions may occur in quite diverse combinations and we have assumed a manner of speech by which we designate as sensory deception certain unusual complex combinations which deviate in certain ways from familiar ones. A child that does not know the qualities of a mirror thinks it has been duped, because the optical impression of the existence of a playmate is not accompanied by the usual other impressions it expects in this situation. The grown-up who already knows that a certain kind of smooth surface favors the formation of such optical impressions with different connotations declares, on his part, that the aerial image known as 'Fata Morgana', with which he is not familiar, is a deception. We all try to orient ourselves in the world, that is to supplement the necessarily incomplete *actual* present observations by experiences collected earlier. A burnt child shies away from the oven because from the optical impression of the oven he infers a tactile impression which in his experience has been disagreeable. Basically he does the same thing that every scholar does in his scientific work. A doctor, for instance, judges the state of a patient by the number of red blood corpuscles, or rather by reading a scale on a certain apparatus; a philologist is led to make far-reaching statements about some historical event upon finding a certain kind of linguistic form in a document. Each of these inferences is basically of the same order, each puts an immediate observation into a familiar context, each has some degree of certainty, none is absolutely reliable. There is no difference of principle between the so-called natural and the humanistic sciences. All assertions, in whatever field they are made, are meaningful only in as much as they can be *verified*, that is, if in the last analysis they can be reduced to statements concerning sense perception.

It does not matter, of course, whether or not such an analysis is really performed in every single case, and even less whether it is short and simple. On the contrary, it may be said that the foundation of all higher intellectual life, in fact, is nothing more than this: we can state in one phrase, containing few words, thoughts which would require volumes to be completely reduced to elements of experience (*Erlebniselemente*). If the chemist asserts that all bodies are composed of a small number of different elements, we do not ask him first to analyse everything near to us. We are satisfied if, on the basis of his thesis, he produces practical and useful results. Similarly, by proposing that it is possible to reduce all statements, no matter how complicated, to elements of experience and observation, Mach's analysis of knowledge furnishes the basis for constructing a science freed from contradictions and from other intellectual risks.

Our world is too complicated to make it possible and expedient for us to look every moment at everything we encounter from *all* sides. For the person who measures a piece of property to build a house on it, the surface of the earth is a plane; the person who examines the course of a river must consider the unevenness of the surface; for one who undertakes to travel around the world, the earth has certain qualities of a spherical surface; the geodesist who performs measurements of a very wide range and of great exactitude, will find that the earth is not a sphere but more aspherical. Each of these points of view is justified within its own context, none is truer or more real or closer to reality than the next. The earth has in fact some properties of a plane and others of a sphere. Sometimes one kind of property is important to us and sometimes another, that's all. A similar answer is given to the question whether the earth is at rest or whether it turns around its axis. It is at rest if the words 'turns itself' imply subjective phenomena which remind one of a carousel. On the other hand, the assumption of a rotation freed from subjective secondary phenomena enables us to describe observable cosmic phenomena in a simpler and more economical manner, that is, if we are interested in doing so.

In this connection we should note that words and phrases used in everyday language were created solely for the purpose of communicating about the most familiar phenomena. When they later were taken up by science for other purposes, they naturally changed their meaning. 'Up and down' are familiar concepts which suffice only to describe relations in a relatively narrow environment. When we speak on an intellectual level where the earth is a globe, they lose their original meaning. Another example: in economic theory, as well as in ordinary language, we speak of the 'income' of a person who lives in a certain kind of economic community. But the 'income' of a savage, or the 'income' of a nation or even of all mankind are non-concepts unless we assign new linguistic significance to them in keeping with the enlarged realm of phenomena. This is, in fact, done in the so-called exact sciences, where expressions of ordinary language assume different meanings, thanks to arbitrarily assigned artificial definitions which differ from common usage. Often some elements of the scientific language penetrate the language of common usage and create confusion. Many people also try to show how educated they are by using technical terms, and this leads to a blurring of the distinction between the normal and the artificially created word meaning; in time this becomes the source of all sorts of illusory philosophical problems, not to speak of those more or less popular 'philosophers' whose entire activity consists in the juggling of half-understood scientific terms.

Colors which please the healthy eye and which do not exist for the color-blind are not vibrations of the air or anything of the kind, but colors; 'red, blue, yellow' are names for sense impressions, useful for communication in daily life, for the practical description of occurrences in the world of experience of all those who are not physicists and who do not happen to occupy themselves with physical questions. If we want to describe more subtle phenomena, relations which may comprise sensations of light and color together with a wide range of other impressions in our experience, then we do find it necessary to abandon the primitive mode of expression of ordinary language. We find that light behaves in some respects like a wave motion and in other respects like a corpuscular emission, and perhaps one day it will be possible to reconcile the two conceptions which confront each other today. The physicist may then say that he has now found what light 'really' is, and this manner of speech which has pragmatic value for him will be accepted. On the other hand, the physiologist who studies the relation between the sensation of light and certain other processes in our body will arrive at the result that color is a specific biological process in the 'visual substance' (*Sehsubstanz*), and he will be right too. The bitter controversy which Goethe carried on against Newton concerning problems in the theory of colors seems to us to lack any basis whatsoever. As a physicist, Goethe limited himself to exploring an arbitrarily set, very narrow range of phenomena, and within it he wanted to make do with the simplest conceptual expedients. The success of these endeavors could not transcend the limitations of the task. Neither the physicist nor the physiologist can benefit today from such a statements as 'colors are the deeds and sufferings of light', or black is the 'representative of darkness', and more of that sort. But the careful experiments and observations which Goethe made and described lose nothing in importance, and the new trend in the physiology of sense perception, founded by Mach and Hering, could start off in many instances from results of Goethe's observations.

We mentioned before that Mach concerned himself, from the very beginning of his scientific career, with general epistemological questions of physics. But this should not be understood to mean that he did not take an active part in experimental research in physics. We owe to Mach various valuable particular findings in optics, acoustics, and the general theory of waves. He devised a method of investigating air currents which is still much in use, and his observations of flying missiles have not as yet been surpassed. Even more important were his results regarding the physiology of the sense organs. He did a great deal of research on the theory of hearing and on the auditory organ; his achievements preceded those of Helmholtz, and in some respects, went beyond them. He made a particularly thorough study of the physiology of motor sensations (*Bewegungsempfindungen*), and clarified, for the first time, their connection with the labyrinthine nerves. In the matter of sensation of color, Mach suggested a theory of color antagonism which took as its point of departure Goethe's theory of color, a trend further pursued by Hering later on. Even if details of his physiological research on the senses appear today somewhat dated or to have been shown erroneous, there remains a considerable body of pioneering work which would amply suffice to assure its author an honorable place in the history of science.

The best known and the most popular of Mach's theories are certain conclusions drawn from his *Analysis of Sensations* which concern the concepts of 'substance' and the 'individual' – the personal 'I' (or Ego). We already referred to the notion of the thing-in-itself and Mach's position with regard to this typical concept of school philosophy. For Mach, the world in which we live, and in which we experience living, is composed of sensations of color, sound, warmth, pressure, and of impressions of time and space. All these elements – Mach calls them 'sensations', one might call them perhaps by a more neutral term such as is used in photography, 'takes' – appear in the most diverse and ever changing combinations. Some connections prove to be relatively stable and durable, more so than others; they give the impression of being the lasting supports of other rapidly changing appearances, and from this circumstance arises the notion of 'the thing' (*das Dinghafte*), of a body from which one can subtract or to which one can add qualities whilst it never ceases to remain the 'same' body. But nature is neither shell not kernel. It is both, inside and outside, essence and contingency. If we substract from an object all its qualities, what remains is not its 'substance' but nothing, just as if we continue to peel an onion. What we touch is no more real than what we see, and what we see is no more real than what we hear. This hierarchy of the senses is only a very primitive makeshift, useful in certain areas of daily life. If (as some philosophers do unconsciously) we try to maintain this hierarchy beyond the narrow limits of practical concerns, if we want to make it the basis of theoretical designs of the world, we arrive at such absurdities as the 'thing-in-itself', of which the neo-Kantian, Hermann Cohen, said that it is the 'essence of scientific knowledge ... a kind of object ... an unknown but nevertheless real object ... not accessible to experience, but accessible to thought ... which attains validity by virtue of its function as a concept'. According to Mach, notions of body and substance are above all eminently *practical* expedients which help us to orient ourselves in our environment. Their origin goes back to the earliest times of the intellectual development of mankind, to the very beginnings of language as a means of everyday communication. Even today these notions are useful – well nigh indispensable – not only in almost all situations of daily life, but also in large areas of scientific research. But they must not be hypostatized unlimitedly, must not be vested in the mantle of 'absolute' validity, or we might end up in the sort of impasse in which du Bois-Reymond spoke his famous *ignorabimus*. "We shall never be able to explain", the articulate physiologist said, "how our sense-impressions can be *traced back* to the movements of the tiniest parts of bodies, the atoms". – "Certainly not", Mach answered, "for the sense impressions are the *primary* factors which in their combinations give us the world." We form concepts, among them the concept of the body which is not at all simple, so that we may be able to orient ourselves in the world. To explain something means to trace something complicated back to something simpler, and not the other way around. 'Body' or 'thing' is a thought symbol and the linguistic expression of a *relatively* stable combination of tactile and visual impressions linked to certain impressions of time and space.

Many a person who will go along with the devaluation of the concept of 'thing' shies away from the further consequence which no longer allows an absolute place – as a last court of appeal in matters of knowledge – to the Ego, the concept of the human individual. One frequently cited phrase from the *Analysis of Sensations*, which many have disputed and others have quoted in a tone of slight horror, is the sentence: "The Ego cannot be saved." What is meant by that? Only that what we call the Ego is a totality of sensations continuously flowing, continuously enlarging itself or narrowing, changing in every respect at all times, a thought symbol for a sum total without definite limits and without exact definition of content. We should say 'it thinks' and not 'I think' as Lichtenberg already argued concerning Descartes' "je pense, donc je suis." It is true that no-one in possession of his senses confuses his own hunger, his own pain, and his own joy with the hunger, the pain, the joy of another individual. But how about the Ego in dreamless sleep? Who acts in the state of hypnosis, one's own or another's Ego? Where is the boundary between hypnotic and any other form of transfer of the will to act – a transfer which may take all kinds of forms, from simple persuasion and intellectual deception to criminal use of drugs. These and many other questions will have to be answered by those who contend that there is an Ego *a priori*, a concept of the individual-in-itself independent of linguistic usage and of arbitrary definitions in a scientific terminology. Everybody knows also that consciousness of one's own Ego may disappear partly or entirely in moments of a very elevated mood, at high points of happiness or unhappiness, and that the sense of Ego often undergoes extreme and sudden changes in the course of illness or similar circumstances. All this shows that the concept of a stable unified Ego is merely a practical expedient no matter how valuable it may be in all kinds of situations and for all kinds of purpose. It is useful for a very large range of experiences, but it becomes an obstacle when we deal with particular, unusual situations or unfamiliar problems. In this connection, we recall a remark in Goethe's diary: "I have been thinking about fiction and science. The evil they cause is due to the need experienced by reflective judgment which creates for its use some kind of image and then later constitutes this image as true and objective, so that what was once helpful, becomes harmful and impedes progress." What has been said here about auxiliary concepts in science, is even more valid for the thought-saving devices which are pre-scientific, and not due to the demands of our reflective judgment, but rather to naive instinctual needs. Such pre-scientific notions take root in our thinking through constant use by generations and then appear to us as indispensable and inevitable. We do not propose to ban these notions from areas where their use is justified, but they must be eliminated in fields where they impede insight and retard further progress of knowledge. This is the case, above all, when we deal with problems on the boundary between physiology and psychology, or when we are concerned with fundamental questions of scientific theory and the theory of knowledge. It is amazing to see what happens when people borrow expressions and concepts from everyday language without critically analysing them. Among philosophers, for instance, serious thought was once given to the question: how is it possible that the concept of a large tree can be accommodated in the much smaller brain of a man?

It was to be foreseen that opponents of Mach's anti-metaphysical way of thinking would use as ammunition against his conception of the individual the never-failing argument that morality was being endangered. If the boundaries of the Ego must be viewed as unsettled and not exactly definable, then people will no longer be convinced that each person is responsible for all he does or does not do, and with that the door is opened for all kinds of immorality. While we cannot measure the results of research in terms of whether or not they are suitable to support traditional moral principles, there is no reason to worry about Mach's investigations; their sole content and aim is critical epistemological analysis. A man who sees in the whole world only *one* totality of sensations which in the case of the individual Ego are merely more strongly linked, is from the beginning immune against the excesses of egoistic instincts. He keeps his mind open to an all-encompassing understanding for which 'humanity' is still too restricted an expression. Whenever in his works Mach touches on questions which fall into the area of so-called practical reason and morality, he expresses himself in terms which – although they have nothing in common with the strict rigidity of the categorical imperative – point towards the idea of a "freer and serene conception of life, which precludes disdaining another's Ego and overestimating one's own". How he behaved himself, during his own life as a scientist, he who knew nothing but work on impersonal tasks, is best expressed in the words of his friend and comrade, Popper-Lynkeus, who survived him by only a few years:

"What I once said about Robert Mayer is just as valid for Mach: he was a 'character-thinker'. Most scientists, with greater or lesser talents, possess more or less intellect, perspicacity, wide knowledge or ability to combine, and these qualities appear firmly established in their heads, so that we may say, these scientists have this or that talent. But with individuals like Robert Mayer, Darwin, or Mach we think involuntarily that they are something unusual, namely of specially gifted natures. The results of their thought as they gradually emerge, appear to come not out of their heads, but out of their whole being, similar to certain naive, fervent works of art. Such thinkers show, even by their diction and manner of expression, that they are not merely scholars (of which there are and always were thousands) but original individuals, personalities of which you cannot say that they are especially bright or perspicacious; rather that as scientists they are simple, clear, absolutely pure and conscientious, scientists for the sake of their own satisfaction rather than in order to shine before the world."[6]

Mach always rejected for himself the name of philosopher: "There is no such thing as a Machian philosophy, but at most a scientific methodology and a psychology of knowledge, and both are, like all scientific theories, preliminary, incomplete efforts." If Mach had not thought exclusively of the German usage of the word 'philosophy', he might have rejected the term with less vehemence or not at all. In France, 'positive' philosophy, or as they prefer to say today *philosophie scientifique*, was a living concept already before Condillac and Comte. Mach fits undoubtedly into this category. The question of category, however, is not important, rather that he acted in fact like a philosopher, at least if we believe that the task of philosophy is to elucidate the facts of life or science, to explore their meaning and to investigate their complexity. Mach himself said: "I would want my point of view in physics to be one that I would not have to abandon at once when looking into another area of science." Today we would certainly say that the only legitimate concept of a philosopher is that of a man who searches for a unified and consistent conception of the world, which reaches beyond the boundaries of special branches of science.

There were some men among Mach's contemporaries, even among academic philosophers, who expressed similar views, especially Richard Avenarius. His *Empirio-criticism* took the form of a complete philosophical system and the title of his book *Kritik der reinen Erfahrung* (*Critique of Pure Experience*) clearly alludes to classical models. Avenarius and Schuppe (with his philosophy of immanence) agreed with Mach in many questions, especially concerning the relation between physical and psychological elements, and in their view of the economical nature of research; but none of them had much lasting influence[7]. In general, academic philosophy criticized and rejected Mach's work but felt nevertheless that this man was an adversary to be taken seriously. Among the numerous polemics and monographs published on the subject, we want to mention two, the work of R. Hönigswald which culminates in the proof that Mach's assertions are irreconcilable with Kant's point of view, and the interesting dissertation of Robert Musil [later the philosophical novelist – Ed.] written under the influence of Riehl, in which Musil tries to show, on the basis of traditional views, that Mach's theories contain contradictions.

Mach's positivism is distinguished from the older positivism of Comte in that Mach attributes the same value to psychical as to physical facts. Mach, moreover, was not in the least inclined to make apodictic statements. Nothing is more foreign to him than the tendency towards absolutism which finally disfigured both the philosophical and the human image of Comte. Mach's work was nearly finished – only the volume *Knowledge and Error* (*Erkenntnis und Irrtum*) and his posthumously published work *Optics* appeared later – when, in the country where Comte's spirit still dominated intellectual pursuits, the mathematician Henri Poincaré began his philosophical writings with an immediate wide response. Essentially in agreement with Mach, Poincaré propounded a similar view of scientific theory which, however, in the form of the so-called 'conventionalism' especially pointed up one particular aspect of the problem. The French public generally considers Poincaré the founder of the *philosophie scientifique*, although Mach, in fact, had proposed almost all of Poincaré's ideas earlier, in a more elaborate and more consistent form. In any case, today a large part of French science is today founded on these doctrines; they had been accepted earlier in England and America.

We should not overlook one essential difference that separates Mach and all those near to him who demand freedom from metaphysics, from various tendencies usually designated as 'materialism' by academic philosophers. Büchner with his 'materialism', Haeckel with his 'monism' placed great emphasis on the claim that they possessed the *final solution* to all of the world's problems. To them, whatever was the last, still incomplete, findings in the natural sciences appeared as the answer to all possible questions, and in this they were not unlike academic philosophers who used to dogmatize on the level of earlier, seemingly well assured, scientific knowledge. By contrast, Mach characteristically sought to re-adjust the claims of science to their proper proportions, and, much more they seemed unsolved and undecided to him than to most philosophers. All that science and philosophy had so far accomplished and were able to accomplish, he considered an ephemeral product compared with the 'naive realism' of daily life, expressed in everyday language, which is the result of thousands of years of evolution. According to Mach, contemporary science "does not claim to be a finished view of the world; rather, it is aware of working towards a future view of the world (a future *Weltanschauung*). The highest philosophy of the scientist consists, in fact, in tolerating an incomplete view of the world, and to prefer it to a seemingly finished but actually insufficient theory."

The direct continuation of Mach's doctrines fell to a group of scholars whose ideological center was Vienna where Mach had worked earlier. They were called the 'Vienna Circle'. They gathered around Moritz Schlick who was also Mach's successor as the Vienna professor of philosophy and who unfortunately died all too suddenly – among them are mathematicians, physicists, philosophers and sociologists (the best known of them are Rudolf Carnap, Philipp Frank, Hans Hahn, Karl Menger, and Otto Neurath. Like Mach, Schlick was a physicist by academic training. He first turned toward academic philosophy and only later, a little while before he received the call to Vienna, 1922, did he turn towards the frontier between physics and theory of knowledge. His decidedly anti-metaphysical attitude is what ties him and his school so closely to Mach. But we should not fail to see that the Vienna Circle went beyond the framework created by Mach and that it accepted a creative new idea coming, so to speak, from the opposite side. Mach had seen his own task in defending the *empirical* side of the natural sciences against the unjustified claims of the apriorists and the absolutists; and there is no doubt that, in the process, mathematical and logical aspects of scientific construction were neglected. Not long before he died, he said, in the preface to the seventh edition of his *Mechanics*: "Both the empirical and the logical aspects of mechanics require investigation. I think that this is made perfectly clear in my book, even though my own work, for very good reasons, deals mainly with the empirical aspect." The logical problems involved in building up the exact sciences began to interest a number of scholars from the middle of the 19th century on, among them especially the mathematician Gottlob Frege. His work was continued and further developed in the great work *Principia mathematica* by Whitehead and Russell, with which Hans Hahn acquainted the Vienna Circle. At the same time, Ludwig Wittgenstein, who also came from Vienna, supplied a fundamental clarification of the relation of language, logic, and science in his *Tractatus logico-philosophicus* of 1922. Schlick became an enthusiastic disciple of Wittgenstein. The characteristic brand of thought of the Vienna Circle, which liked to call its point of view "logical empiricism', evolved by combining the empiricist views of Mach with the linguistic-logical theory suggested by Wittgenstein. Among the most significant works of the circle were Carnap's *Logische Syntax der Sprache* (*The Logical Syntax of Language*), Frank's *Kausalgesetz und seine Grenzen* (*The Law of Causality and its Limits*), as well as various sociological publications by Neurath.

But the true effect of Mach's doctrines cannot be measured by the number and the success of those who, inspired by his writings, tried to continue his work in the same or a similar direction. In the history of intellectual life, we often encounter the emergence of ideas which prevail gradually and almost unnoticed, and which slowly transform man's way of thinking. Such ideas, often long in preparation, are, at last, adequately expressed by the right person; and eventually they became part of the anonymous common body of thought. Thus, there is hardly any physicist today whose attitude towards the fundamental questions of science, sometimes unconscious, has not been determined by the fact that Mach shook the foundations of dogmatic belief and concretely elucidated certain fundamental concepts of physics. What educated person, moreover, provided he is not a professional metaphysician, would today reject the view that the goal of all science is to present what can be experienced in an orderly manner (i.e., in a manner which exercises economy of thought) and to be silent about what can *not* be experienced (what, in principle, cannot be observed)? Details do not matter; what is important is the one idea of expounding a realm of thought, free from metaphysics, which encompasses all of science. There is no doubt that this idea is gaining ever wider acceptance. Mach never cared at all whether or not it was *his* name which was connected with thoughts that originated with him. This attitude corresponded to his scientific conception of the individual, that in the living Ego personal and impersonal elements co-exist; and that only the latter may endure in some measure beyond the earthly life of a person. The greatest personal modesty and reserve were characteristic of his way of life and with that of the style of his writings. Just as his voice was never loud, just as his movements were never abrupt, just as there was always a calm look in his eyes, so also each line he wrote is imbued with that solemn tranquillity which alone is able to penetrate vast spaces and times. "Thus quiet men have more profoundly, more fruitfully, reached into the future than noisy ones. To the quiet in the land, space and time often became the vault of a language which made their voices loudly heard abroad."[8]

BIOGRAPHICAL NOTE

Ernst Mach was born on the 18th of February 1838 in Turas, Moravia. At first he received instruction from his father. Later he attended secondary school, the *Piaristen Gymnasium* in Kremsier. From 1854 to 1859, he studied physics at the University of Vienna. Here he qualified as a lecturer (*Privatdozent*) in physics in 1861. In 1864 he was called to the University of Graz as a full professor in mathematics, and in 1865, he was named to the chair of physics at the same university. From 1867 to 1895, Mach was professor of experimental physics at the University of Prague. In 1895, he accepted a professorship in philosophy in Vienna which, in accordance with his express wish, was called "professorship for philosophy, especially of the history and the theory of the inductive sciences". In 1898, he fell gravely ill and in 1901, he retired, and at the same time became a member of the Austrian House of Peers (*Herrenhaus*). From 1913 to his death on February 19, 1916, he lived near Munich, at the house of his son, Dr. Ludwig Mach.

REFERENCES

* Translated by Inge Marcuse from *Einheitswissenschaft*, No. 7: Van Stockum, 's-Gravenhage, Holland, 1938, with the kind support of the Institute for the Unity of Science. The English text was edited by Hilda Geiringer Mises and R. S. Cohen. Permission to publish a translation of the article was kindly granted by Mrs. Mises, to whom we are grateful – Ed.

1 Goethe's essay *Bedenken und Ergebung* (*Doubt and Surrender*) is contained in the first volume of his book *On Morphology* (2nd part) 1820.

2 Compare the evaluation of Kant's position, given here, with Nietzsche's statement (*Fröhliche Wissenschaft*, 193) also mentioned by Philipp Frank (*Das Kausalgesetz und seine Grenzen*, p. 281): "Kant wanted to prove to *everybody* in a manner that would antagonize *everybody* that *everybody* was right: – that was the secret joke of this mind. He wrote against the scholar, favoring the prejudices of the common man, but he wrote for scholars and not for the common man."

3 The first passage is from *Analyse der Empfindungen* (*Analysis of Sensations*), 6th German edition, p. 24, the second, *op. cit.*, p. 299. Compare also the following para-

graph from Mach's *Leitgedanken* (*Guiding thoughts*), p. 12: "I must speak of one more influence which helped to determine the direction of my thought. It is very nearly the first one, but for special reasons I mention it last. Already in 1853, in my very early youth, the naive realism of my view of the world was violently shaken by Kant's *Prolegomena*. When I realized two or three years later that the 'thing in itself' was nothing but idle illusion, I returned to Berkeley's point of view which is latently inherent in Kant. An idealistic mood was, however, hard to reconcile with the study of physics. The torment grew worse when I became acquainted with Herbart's mathematical psychology and Fechner's psycho-physics which linked intimately what was acceptable and what was not acceptable. Fortunately or unfortunately I did not have the means to undertake research in physics when I had finished my studies at the university, and I was thus forced to enter the area of physiology of the senses. It is here, where I could observe my sensations and, at the same time their conditioning within an environment, that I arrived at what I consider to be a natural conception of the world free from speculative, metaphysical additives. The dislike of metaphysics instilled in me by Kant's, Herbart's and Fechner's analyses led me back to a point of view very close to that of Hume."

[4] The remark made by Goethe on the 18th of June 1826 to Chancellor F. v. Müller was noted down by the latter and has been reprinted in Fl. v. Biedermann, *Goethe's Gespräche* [*Conversations with Goethe*], vol. III, Leipzig (1910), p. 273.

[5] From 'Geometry and Experience' in *Sidelights on Relativity*, Methuen, London, 1922, p. 28. [The original lecture was delivered to the Prussian Academy of Sciences in Berlin, January 27, 1921. In fact, Einstein was not present at his Nobel presentation and hence delivered no address of acceptance. – Ed. note.]

[6] The obituary for Mach, written by Josef Popper-Lynkeus, appeared in the *Vossische Zeitung*, Berlin 1916. p. 31.

[7] Mach expresses his views on his "Relations to R. Avenarius and other scientists" in the *Analysis of Sensations*, 6th German edition, pp. 38–40. If we look at Avenarius' *Kritik der reinen Erfahrung*, 2nd ed., Leipzig 1907, we can find hardly any similarity between his own manner of presentation and the clear and unambiguous style of Mach, and it is difficult to see that they both agreed on the main ideas. Avenarius' doctrinaire manner of presentation, his innumerable neologisms, are probably the reason why his writings had very little success.

[8] Citation from Jean Paul's fictional biography: *Das Leben Fibels*.

APPENDIX D

ERNST MACH: BIOGRAPHICAL DATA*

Compiled by Otto Blüh

I. PERSONAL AND FAMILY[1]

1838	Born 18, February, in Turas, Moravia, Czechoslovakia, as son of Johann Mach (1805–1879), high-school teacher, and Josefine neé Lanhaus (1813–1869). Sisters: Octavia, m. Josef Engelhart, and Marie (d. 1928). Marie wrote *Erinnerungen einer Erzieherin*, W. Braumüller, Vienna, 2nd ed., 1913. With a Foreword by Ernst Mach (1911).
1867	Married to Louise Marussig (d. 1921). Sons: Ludwig (1868–1951), M.D., Scientist and Inventor; Heinrich (1874–1894), Student of physics at Göttingen; Felix (1879–1933), Landscape painter; Victor (1881–1942), Instrument designer. Daughter: Caroline (1873–1965), married to Dr. Anton Lederer; Son: Dr. Anton Ernst Lederer (b. 1895), Physicist; in U.S.A. since 1923.
1897	Suffered paralytic stroke.
1912	Moved with his wife to the home of his son Dr. Ludwig Mach in Vaterstetten, Bavaria.
1916	Died 19 February, in Vaterstetten.

II. PROFESSIONAL

1856	Began study at Vienna University.
1860	Doctor of Philosophy degree.
1861	*Privat-dozent* of Physics.
1864	Professor of Mathematics and Physics at Graz University.
1867	Professor of Physics and Director of the Department of Physics at Prague University.
1872/73	Dean of the Philosophical Faculty.
1876	Title of *Regierungsrat* (State Councillor).

1879/80	Rector of Prague University.
1879/80	Physics Department moved into the new 'Science Institute'.
1882/83	Prague University divided into two independent Universities (Czech and German).
1883/84	Rector of the German Karl-Ferdinands University of Prague.
1895	Professor of Philosophy, particularly for the History and Theory of the Inductive Sciences, Vienna University.
1896	Title of *Hofrat* (Court Councillor); Secretary of the Mathematical-Scientific Division of the Imperial Academy of Vienna.
1898	Leave of absence for 6 months, after suffered partial paralysis.
1899	Reduced teaching and examination duties.
1900	Retired 1 May.
1901	Appointed Member of the *Herrenhaus* (Upper Chamber of the Austrian Parliament.)

III. HONORS

Doctor of Medicine, hon. causa, University of Tübingen.

Corresponding Member of the Leopoldine-Caroline Academy of Halle, the Bavarian Academy in Münich, the Academy of Göttingen, and the Instituto de Coimbra. Recipient of several medals.

IV. MACH'S COLLEAGUES IN PRAGUE[2]

V. Pierre was the immediate predecessor of Mach; he became professor at the *Technische Hochschule* in Vienna.

Vinzenz Dvořak offered 1874–1876 courses on the history of physics.

Ewald Hering (1834–1918), physiologist, succeeded Purkyně in 1870. He went to Leipzig in 1895.

Gustav Jaumann and Ottokar Tumlirz were assistants of Mach and later became respectively professors at the *Technische Hochschule* of Brünn (Brno) and the University of Czernowitz.

Ernst Lecher (1856–1926) was the successor of Mach (1895 to 1909).

Ferdinand Lippich (1838–1913), was named professor of mathematical physics in 1872; he retired 1908. His successor was Albert Einstein, who was in Prague from 1910 to 1912.

Johannes Evangelista Purkyně (Purkinye) (1787–1869), physiologist.

August Seydler, also a docent under Mach, was the first professor of mathematical physics at the Czech University.

Vinzenz Strouhal studied under Mach and Friedrich Kohlrausch in Würzburg, became docent of physics in Prague 1878, and the first professor of experimental physics at the Czech University of Prague, 1882.

REFERENCES

* [Cf. the biographical summary at the end of the essay by Mises (Appendix, p. 269) for several minor emendations – Ed.]

[1] I wish to express my thanks to Dr. A. E. Lederer for the information contained in this summary.

[2] I am greatly indebted to Dr. Julius Kroczek of Prague for collecting this information.

ERNST MACH BIBLIOGRAPHY

Compiled by O. Blüh and W. F. Merzkirch

(Adopted from Joachim Thiele, 'Ernst Mach-Bibliographie', *Centaurus* **8** (1963), 189–237.)

I. WORKS OF ERNST MACH IN ORIGINAL EDITIONS

Prepared by W. F. Merzkirch

This short bibliography contains only the original German titles of Mach's publications. Not listed are: All translations into foreign languages; preliminary communications which have been published later in final form; some short notes to *EJ*, *MU* and *ZphU*; some preliminary lectures which have later been published in *Populär-wissenschaftliche Vorlesungen*; and the text-books written for high-schools.*

The titles marked with * are also printed in *Populär-wissenschaftliche Vorlesungen*. The publications appearing in *SW* and marked with ** have later been reprinted in *AP*.

For a more detailed information the reader is referred to the Ernst-Mach-Bibliography by Joachim Thiele (in *Centaurus* **8**, 1963, 189–237).

Abbreviations:

AP — *Annalen der Physik*, Leipzig
CR — *Carls Repertorium*, München
EJ — *Jahrbuch für Photographie und Reproductionstechnik* (ed. by J. M. Eder), Halle
MU — *Untersuchungen zur Naturlehre des Menschen und der Tiere* (ed. by J. Maleschott), Giessen
SW — *Sitzungsbericht der Kaiserlichen Akademie der Wissenschaften, Mathematisch-Naturwissenschaftliche Classe*, Wien
ZMP — *Zeitschrift für Mathematik und Physik*, Leipzig
ZphU — *Zeitschrift für den physikalischen und chemischen Unterricht*, Berlin

1859

Blaserna-Mach-Peterin, 'Über elektrische Entladung und Induction', *SW* 37, 477–524.

1860

'Über die Änderung des Tones und der Farbe durch Bewegung', *SW* **41, 543–560.

1861

'Über das Sehen von Lagen und Winkeln durch die Bewegung des Auges', *SW* **43**, 215–224.

* Cf. 'List of Mach's Educational Publications' by Otto Blüh, this volume, pp. 20–21.

'Über die Controverse zwischen Doppler und Petzval bezüglich der Änderung des Tones und der Farbe durch Bewegung', *ZMP* **6**, 120–126.

1862

'Über die Molecularwirkung der Flüssigkeiten', *SW* **46**, 125–134.
'Zur Theorie der Pulswellenzeichner', *SW* **46**, 157–174.
'Über die Änderung von Ton und Farbe durch Bewegung', *AP* **116**, 333–338.

1863

Compendium der Physik für Mediciner, Wien 1863.
'Über die Gesetze des Mitschwingens', *SW* **47**, 33–48.
'Über eine neue Einrichtung des Pulswellenzeichners', *SW* **47**, 53–56.
'Zur Theorie des Gehörorgans', *SW* **48**, 283–300.
'Vorlesungen über Psychophysik', *Oesterreichische Zeitschrift für praktische Heilkunde* **9**.

1864

'Über einige der physiologischen Akustik angehörige Erscheinungen', *SW* **50**, 342–362.

1865

'Untersuchungen über den Zeitsinn des Ohres', *SW* **51**, 133–150.
'Bemerkungen über die Accomodation des Ohres', *SW* **51**, 343–346.
'Über die Wirkung der räumlichen Vertheilung des Lichtreizes auf die Netzhaut, part I', *SW* **52**, 303–322.
'Über Flüssigkeiten, welche suspendirte Körperchen enthalten', *AP* **126**, 324–330.
'Bemerkungen über den Raumsinn des Ohres', *AP* **126**, 331–333.
'Über die anschauliche Darstellung einiger Lehren der musikalischen Akustik', *ZMP* **10**, 425–428.
'Bemerkungen über intermittirende Lichtreize', *Archiv für Anatomie, Physiologie und wissenschaftliche Medicin*.

1866

Einleitung in die Helmholtz'sche Musiktheorie. Populär für Musiker dargestellt, Graz 1866.

'Über wissenschaftliche Anwendungen der Photographie und Stereoskopie', *SW* **54**, 123–126.
'Über den physiologischen Effect räumlich vertheilter Lichtreize, part II', *SW* **54**, 131–144.
'Über die physiologische Wirkung räumlich vertheilter Lichtreize, part III', *SW* **54**, 393–408.
'Über eine Vorrichtung zur mechanisch-graphischen Darstellung der Schwingungscurven', *AP* **129**, 464–466.
'Über die Wellen mit Flüssigkeit gefüllter elastischer Röhren', *MU* **10**, 71–74.
'Bemerkungen über die Entwicklung der Raumvorstellungen', *Fichtes Zeitschrift für Philosophie* **49**, 227–232.

1867

**Zwei populäre Vorträge über Optik*, Graz 1867.
'Über eine Longitudinalwellenmaschine', *AP* **132**, 174–176.

1868

'Über die physiologische Wirkung räumlich vertheilter Lichtreize, part IV', *SW* **57**, 11–19.
'Beobachtungen über monoculare Stereoskopie', *SW* **58**, 731–736.
'Einfache Demonstration des Huyghens'schen Princips', *AP* **134**, 310–311.
'Einfache Demonstration der Schwingungsgesetze gestrichener Saiten', *AP* **134**, 311–312.
'Über die Definition des Masse', *CR* **4**, 355–359.
'Über die Versinnlichung einiger Sätze der Mechanik', *CR* **4**, 359–361.
'Über die Versinnlichung der Poinsot'schen Drehungstheorie', *CR* **4**, 361.
'Über die Abhängigkeit der Netzhautstellen voneinander', *Vierteljahresschrift für Psychiatrie*, Neuwied **2**.

1870

'Mittheilungen über einfache Vorlesungsversuche', *CR* **6**, 8–12.

1871

'Optische Vorlesungsversuche', *CR* **7**, 261–264.

'Die mechanische Nachahmung des Fermat'schen Brechungsgesetzes', *CR* **7**, 375–377.

'Eine Bemerkung über den zweiten Hauptsatz der mechanischen Wärmetheorie', *Lotos* **21**, 17–18.

'Über die physikalische Bedeutung der Gesetze der Symmetrie', *Lotos* **21**, 139–147.

1872

Die Geschichte und die Wurzel des Satzes von der Erhaltung der Arbeit, Prag 1872; 2nd edition: 1909.

**Die Gestalten der Flüssigkeit. Die Symmetrie*, Prag 1872.

'Über die stroboskopische Bestimmung der Tonhöhe', *SW* **66, 267–274.

Mach-Kessel, 'Die Funktion der Trommelhöhle und der Tuba Eustachii', *SW* **66**, 329–336.

Mach-Kessel, 'Versuche über die Accomodation des Ohres', *SW* **66**, 337–343.

'Über die temporäre Doppelbrechung der Körper durch einseitigen Druck', *AP* **146**, 313–316.

'Spectrale Untersuchung eines longitudinal tönenden Glasstabes', *AP* **146**, 316–317.

1873

Optisch-akustische Versuche. Die spectrale und stroboskopische Untersuchung tönender Körper, Prag 1873.

Mach-Fischer, 'Die Reflexion und Brechung des Schalles', *SW* **67, 81–88.

'Über die Stefan'schen Nebenringe am Newton'schen Farbenglas und einige verwandte Interferenzerscheinungen', *SW* **67**, 371–381.

'Physikalische Versuche über den Gleichgewichtssinn des Menschen, part I', *SW* **68**, 124–140.

'Zur Geschichte des Arbeitsbegriffs', *SW* **68**, 479–488.

'Resultate einer Untersuchung zur Geschichte der Physik', *Lotos* **23**, 189–191.

1874

'Versuche über den Gleichgewichtssinn, part II', *SW* **69**, 121–135.

'Über den Gleichgewichtssinn, part III', *SW* **69**, 44–51.

Mach-Kessel, 'Beiträge zur Topographie und Mechanik des Mittelohres', *SW* **69**, 221–243.

1875

Grundlinien der Lehre von den Bewegungsempfindungen, Leipzig 1875.

Mach-Wosyka, 'Über einige mechanische Wirkungen des elektrischen Funkens', *SW* **72, 44–52.

Mach-Rosicky, 'Über eine neue Form der Fresnel-Arago'schen Interferenzversuche', *SW* **72**, 197–212.

Mach-Merten, 'Bemerkungen über die Änderung der Lichtgeschwindigkeit im Quarz durch Druck', *SW* **72**, 315–328.

'Über einen Polarisationsapparat mit rotirendem Zerleger', *AP* **156**, 169–172.

'Bemerkungen über die Funktion der Ohrmuschel', *Archiv für Ohrenheilkunde* **9**, 72–76.

1877

Mach-Sommer, 'Über die Fortpflanzungsgeschwindigkeit von Explosionsschallwellen', *SW* **75**, 101–130.

1878

Mach-Tumlirz-Kögler, 'Über die Fortpflanzungsgeschwindigkeit der Funkenwellen', *SW* **77**, 7–32.

'Neue Versuche zur Prüfung der Doppler'schen Theorie der Ton- und Farbenänderung durch Bewegung', *SW* **77**, 299–310.

'Über den Verlauf der Funkenwellen in der Ebene und im Raume', *SW* 77, 819–838.

Mach-Gruss, 'Optische Untersuchung der Funkenwellen', *SW* **78**, 476–480.

Mach-Weltrubsky, 'Über die Formen der Funkenwellen', *SW* **78**, 551–560.

Mach-Doubrava, 'Über die elektrische Durchbrechung des Glases', *SW* **78, 729–732.

1879

Mach-Doubrava, 'Beobachtungen über die Unterschiede der beiden elektrischen Zustände', *SW* **80, 331–345.

Mach-Simonides, 'Weitere Untersuchung der Funkenwellen', *SW* **80**, 476–486.

1882

**Die ökonomische Natur der physikalischen Forschung*, Wien 1882.

'Über Herrn A. Guébhard's Darstellung der Aequipotentialcurven', *SW* **86, 8–14.

1883

Die Mechanik in ihrer Entwickelung historisch-kritisch dargestellt, Leipzig 1883; further German editions: 1888, 1897, 1901, 1904, 1908, 1912, 1921, 1933.

'Versuche und Bemerkungen über das Blitzableitersystem des Herrn Melsens', *SW* **87**, 632–639.

1884

**Über Umbildung und Anpassung im naturwissenschaftlichen Denken*, Wien 1884.

*'Über die Grundbegriffe der Elektrostatik', *Lotos* **33**, 90–112.

1885

Mach-Arbes, 'Einige Versuche über totale Reflexion und anomale Dispersion', *SW* **92, 416–426.

Mach-Wentzel, 'Ein Beitrag zur Mechanik der Explosionen', *SW* **92, 625–638.

'Zur Analyse der Tonempfindungen', *SW* **92**, 1283–1289.

1886

Beiträge zur Analyse der Empfindungen, Jena 1886; further German editions: 1900, 1902, 1903, 1906, 1911, 1918, 1919, 1922.

**Der relative Bildungswert der philologischen und der mathematischen-*

naturwissenschaftlichen Unterrichtsfächer der höheren Schulen, Prag 1886.

1887

Mach-Salcher, 'Photographische Fixirung der durch Projectile in der Luft eingeleiteten Vorgänge', *SW* **95, 764–778.

'Über den Unterricht in der Wärmelehre', *ZphU* **1**, 3–7.

1888

'Über die Fortpflanzungsgeschwindigkeit des durch scharfe Schüsse erregten Schalles', *SW* **97**, 1045–1052.

'Über eine Lichtquelle zum Photographiren nach der Schlieren-methode', *EJ* **2**, 284.

'Ergebnisse der Momentphotographie', *EJ* **2**, 287–290.

'Über die Anordnung von quantitativen Schulversuchen', *ZphU* **1**, 197–199.

'Die experimentelle Darstellung der Linsenabweichungen', *ZphU* **2**, 52–55.

1889

Mach-Salcher, 'Über die in Pola und Meppen angestellten ballistisch-photographischen Versuche', *SW* **98**, 41–50.

'Über die Schallgeschwindigkeit beim scharfen Schuss nach von dem Krupp'schen Etablissement angestellten Versuchen', *SW* **98**, 1257–1276.

Mach-Salcher, 'Optische Untersuchung der Luftstrahlen', *SW* **98, 1303–1309.

Mach-Mach (Ludwig), 'Weitere ballistisch-photographische Versuche', *SW* **98**, 1310–1326.

Mach-Mach (Ludwig), 'Über longitudinale fortschreitende Wellen im Glase', *SW* **98**, 1327–1332.

Mach-Mach (Ludwig), 'Über die Interferenz der Schallwellen von grosser Excursion', *SW* **98**, 1333–1336.

1890

'Sphärische Concavspiegel zur Photographie mittels des Schlieren-apparates', *EJ* **4**, 108–109.

'Über das psychologische und logische Moment im naturwissenschaftlichen Unterricht', *ZphU* **4**, 1–5.

1891

Leitfaden der Physik für Studierende, Prag & Wien; Leipzig 1891; 2nd edition: Leipzig 1891.
'Über weitere Fortschritte in der Momentphotographie', *EJ* **5**, 166–167.

1892

'Ergänzungen zu den Mitteilungen über Projectile', *SW* **101**, 977–983.
'Zur Geschichte und Kritik des Carnot'schen Wärmegesetzes', *SW* **101**, 1589–1612.
'Über eine elementare Darstellung der Fraunhofer'schen Beugungserscheinungen', *ZphU* **5**, 225–229.

1893

Mach-Doss, 'Bemerkungen zu den Theorien der Schallphänomene bei Meteoritenfällen', *SW* **102**, 248–252.

1894

'Über ein Princip der Verstärkung unterexponierter Bilder', *EJ* **8**, 152–153.
'Über die rasche Ermittlung der richtigen Expositionszeit', *EJ* **8**, 154–156.
*'Über das Princip der Vergleichung in der Physik', *Verhandlungen der Gesellschaft Deutscher Naturforscher und Ärzte*.
'Einfache Versuche über strahlende Wärme', *ZphU* **7**, 113–116.

1896

Die Principien der Wärmelehre. Historisch-kritisch entwickelt, Leipzig 1896; further German editions: 1900, 1919.
Populär-wissenschaftliche Vorlesungen, Leipzig 1896 (15 lectures);

further German editions: 1897 (15 lectures), 1903 (19), 1910 (26), 1923 (33).
'Durchsicht-Stereoskopbilder mit Roentgenstrahlen', *Zeitschrift für Elektrotechnik*, Wien **14**, 259–261.
'Über Gedankenexperimente', *ZphU* **10**, 1–5.

1897

'Bemerkungen über die historische Entwicklung der Optik', *ZphU* **11**, 3–8.

1902

'Die Ähnlichkeit und die Analogie als Leitmotiv der Forschung', *Annalen der Naturphilosophie*, Leipzig **1**, 5–14.

1904

'Objektive Darstellung der Interferenz des polarisierten Lichtes', in *Festschrift Ludwig Boltzmann*, Leipzig 1904, pp. 441–447.
Mach-Mach (Ludwig), 'Versuche über Totalreflexion und deren Anwendung', *SW* **113**, 1219–1230.

1905

Erkenntnis und Irrtum. Skizzen zur Psychologie der Forschung, Leipzig 1905; further German editions: 1906, 1917, 1920, 1926.

1906

'Über den Einfluss räumlich und zeitlich variierender Lichtreize auf die Gesichtswahrnehmung', *SW* **115**, 633–648.
'Über die Phasenänderung des Lichtes durch Reflexion', in *Festschrift Adolf Lieben*, Leipzig 1906, pp. 291–296.

1907

'Die Phasenverschiebung durch Reflexion an den Jamin'schen Platten', *SW* **116**, 997–1000.

1910

'Die Leitgedanken meiner naturwissenschaftlichen Erkenntnislehre', *Physikalische Zeitschrift* **11**, 599–606.

*'Eine Betrachtung über Zeit and Raum', in *Das Wissen für Alle*, Wien 1910.

'Sinnliche Elemente und naturwissenschaftliche Begriffe', *Pflüger's Archiv für Physiologie* **136**, 263–274.

'Die Organisierung der Intelligenz', *Neue Freie Presse*, Wien, July 24.

1911

*'Psychisches und organisches Leben', *Österreichische Rundschau*, Wien **29**, 22–31.

*'Allerlei Erfinder und Denker', *Naturwissenschaftliche Wochenschrift* **26**, 497–501.

1912

*'Das Paradoxe, das Wunderbare und das Gespenstische', *Kosmos* **9**, 17–20.

*'Psychische Tätigkeit, insbesondere Phantasie, bei Mensch und Tier', *Kosmos* **9**, 121–125.

1915

Kultur und Mechanik, Stuttgart 1915.

'Einige Experimente über Interferenz, insbesondere über complementärfarbige Interferenzringe', in *Festschrift Wilhelm Jerusalem*, Wien 1915, pp. 154–173.

1916

*'Einige vergleichende tier- und menschenpsychologische Skizzen', *Naturwissenschaftliche Wochenschrift* **31**, 241–247.

1921

Die Principien der physikalischen Optik. Historisch und erkenntnispsychologisch entwickelt, Leipzig 1921.

II. ENGLISH TRANSLATIONS OF BOOKS BY ERNST MACH

The Science of Mechanics. A critical and historical exposition of its principles (transl. from the 2nd German edition by Thomas J. McCormack), The Open Court Publ. Co., Chicago, 1893; Watts & Co., London, 1893.
2nd rev. and enl. ed., The Open Court Publ. Co., Chicago, 1902.
3rd ed., The Open Court Publ. Co., Chicago, 1907.
Supplement to the 3rd English translation, containing the author's addition to the 7th German ed. (transl. by Ph. E. B. Jourdain), The Open Court Publ. Co., Chicago, 1915.
4th ed., The Open Court Publ. Co., Chicago, 1919.
5th ed. La Salle, Ill., The Open Court Publ. Co., London, 1942. (Containing additions and alterations up to the 9th German ed.)
6th edition. New Introduction by Karl Menger, La Salle, Ill., The Open Court Publ. Co. 1960.

Popular Scientific Lectures (Transl. by Thomas J. McCormack), The Open Court Publ. Co., Chicago, 1895 (12 lectures).
2nd ed., rev. and enl. The Open Court Publ. Co., Chicago, 1897 (13 lectures).
3rd ed., rev. and enl., The Open Court Publ. Co., Chicago, K. Paul, Trench, Truebner & Co., London, 1898. Reprint of 3rd ed., 1910.

Contributions to the Analysis of the Sensations (transl. by C. M. Williams), The Open Court Publ. Co., Chicago, 1897.

The Analysis of Sensations and the Relation of the Physical to the Psychical (transl. from the 1st German ed. by C. M. Williams, rev. and suppl. from the 5th German ed. by Sidney Waterlow), The Open Court Publ. Co., Chicago and London, 1914. (With a new introduction by Thomas S. Szasz) Dover Publ. New York, 1959.

History and Root of the Principle of the Conservation of Energy (transl. from the German and annotated by Philip E. B. Jourdain), The Open Court Publ. Co., Chicago, 1911.

Space and Geometry in the light of physiological, psychological and physical inquiry. (transl. from the German by Thomas J. McCormack), The Open Court Publ. Co., Chicago, 1906 (Reprinted from *The Monist*, April 1901, July 1902, October 1903).

The Principles of Physical Optics. An historical and philosophical Treatment. (transl. by John S. Anderson and A. F. A. Young), Methuen & Co., London, 1926; Dover Publ., New York, 1953.

III. ENGLISH TRANSLATIONS OF ARTICLES BY ERNST MACH

In: *Philosophical Magazine*, Ser. 4. (London)

'On the Visual Sensations Produced by Intermittent Excitations of the Retina', **30** (1865), 319.

'On an Arrangement for the Graphical Representation of Curves of Vibration by Means of Mechanism', **33** (1866), 159.

In: *The Open Court. A Weekly Journal devoted to the Religion of Science.* (Chicago)

'Transformation and Adaptation in Scientific Thought', **2** (1888), 1087, 1115.

'The Forms of Liquids', **8** (1894), 3935.

'Symmetry', **8** (1894), 4015.

'The Fibres of Corti', **8** (1894), 4087.

'On the Causes of Harmony', **8** (1894), 4136.

'The Velocity of Light', **8** (1894), 4167.

'Why has Man two Eyes?' **8** (1894), 4175.

'On the Fundamental Concepts of Electrostatics', **8** (1894) 4247, 4255.

'The Economical Character of Physical Research', **8** (1894), 4263, 4271.

'On the Principle of Comparison in Physics', **8** (1894), 4283, 4288.

'On the Relative Educational Value of the Classics and the Mathematico-Physical Sciences in Colleges and High Schools', **8** (1894), 4295, 4308, 4311.

N.B. All the above reprinted in *Popular Scientific Lectures* (1895).

'Names and Numbers', **14** (1900), 37.

'Language' (Its Origin, Development and Significance for the Development of Thought), **14** (1900), 171.

'The Concept', **14** (1900), 348.

'The Notion of a Continuum' (An Essay in the Theory of Science), **14** (1900), 409.

'The Propensity toward the Marvellous', **14** (1900), 539.

'The Theory of Heat', **16** (1902), 641.

'Sketch of the History of Thermometry', **16** (1902), 733; **17** (1903), 26.

'Critique of the Concept of Temperature', **17** (1903), 95, 154.

'On the Determination of High Temperatures', **17** (1903), 297.

N.B. The last 9 articles are translations of chapters from Mach's *Wärmelehre* (*Principles of Heat*) (1896).

In: *The Monist. A Quarterly Magazine Devoted to the Philosophy of Science.* (Chicago)

'The Analysis of Sensations – Antimetaphysical', **1** (1890), 48.
'Facts and Mental Symbols', **2** (1892), 198.
'On the Principle of Conservation of Energy', **5** (1894), 22.
'On the Part Played by Accident in Invention and Discovery', **6** (1896), 161.
'On the Stereoscopic Application of Röntgen's Rays', **6** (1896), 321.
'On Sensations of Orientation', **8** (1897), 79.
'On Physiological, as Distinguished from Geometrical Space', **11** (1906), 321.
'On the Psychology and Natural Development of Geometry', **12** (1902), 481.
'Space and Geometry from the Point of View of Physical Inquiry', **14** (1903), 1. (The German text published in *Erkenntnis und Irrtum*, 1905).
'Inventors I have met', **22** (1912), 230.
'Psychic and Organic Life', **23** (1913), 1.

IV. PUBLICATIONS ON ERNST MACH IN ENGLISH

Blanshard, Brand, *Reason and Analysis* (*In memory of Paul Carus*, 1852–1919), chap. III: 'The Rise of Positivism', Allen and Unwin, Ltd., London, 1962.
Blüh, Otto, 'Ernst Mach as Teacher and Thinker,' *Physics Today* **20** (1967), 32.
Blüh, Otto, 'Ernst Mach as an Historian of Physics', *Centaurus* **13** (1968), 62.
Bridgman, P. W., 'Significance of the Mach Principle', *Am. J. Phys.* **29** (1961), 32.
Bunge, Mario, 'Mach's Criticism of Newtonian Mechanics', *Am. J. Phys.* **34** (1966), 585.
Carus, Paul, 'Feelings and the Elements of Feelings', *The Monist* **1** (1891), 401.
Carus, Paul, 'Professor Mach's Term Sensation', Supplementary to his Controversy with the Editor (Carus), *The Monist* **3** (1893), 298.
Carus, Paul, 'Professor Mach's Philosophy', *The Monist* **16** (1906) 627.

Carus, Paul, 'Professor Mach and his work', *The Monist* **21** (1911), 19.

Cohen, M. R., *Reason and Nature. An Essay on the Meaning of Scientific Method.*, chap. II, Sec. 2., Free Press, Glencoe, Ill., 1953.

Dicke, R. H., 'Cosmology, Mach's Principle and Relativity', *Am. J. Phys.* **31** (1963), 500.

Feld, James W., 'Mach's Principle Revisited', *Laval Theologique et Philosophique* **20** (1964), 35.

Frank, Philipp, 'Ernst Mach. The Centenary of his Birth'. *Erkenntnis. Journal for Unified Science* **7** (1937/38), 247.

Frank, Philipp, 'Einstein, Mach and Logical Positivism', in: *Albert Einstein – Philosopher – Scientist* (ed. by P. A. Schilpp), The Library of Living Philosophers, Inc., Evanston, Ill., vol. **VII**, 1949.

Gardner, Martin, *The Ambidextrous Universe*, chap. 19, p. 189: 'Mach's Shock', Basic Books, Inc. New York, 1964.

Höffding, Harald, *Modern Philosophers* (transl. by Alfred C. Mason), section on Mach, pp. 115–121. Macmillan & Co., London, 1920.

Jourdain, Ph. E. B., 'The Principle of Least Action. Remarks on Some Passages of Mach's *Mechanics*', *The Monist* **22** (1912), 285.

Kleinpeter, Hans, 'On the Monism of Professor Mach', *The Monist* **16** (1906), 161.

Kolakowski, Leszek, *The Alienation of Reason. A History of Positivist Thought.* (Original Polish title: *Positivist Philosophy. From Hume to the Vienna Circle*) (transl. by Norbert Guterman), Doubleday, New York, 1968. On Mach, p. 118ff.

Mises, Richard von, *Positivism. A Study in Human Understanding* (transl. by J. Bernstein and R. G. Newton), Harvard University Press, Cambridge, 1951. Reprint, unchanged, Dover Publ., New York, 1968.

Moon, P. and D. E. Spenser, 'Mach's Principle', *Philosophy of Science* **26** (1959), 125.

Pittenger, H. W., 'Ernst Mach: Biographical Notes', *Science* **150** (1965), 1120.

Ratliff, Floyd, *Mach Bands: Quantitative Studies on Neural Networks in the Retina*, Holden-Day, Inc., San Francisco, 1965.

Seaman, Francis, 'Mach's Rejection of Atomism', *J. Hist. Ideas* **29** (1968), 381.

Titchener, E. B., 'Mach's Lectures on Psychophysics', *Am. J. Psych.* **33** (1922), 213.

Weinberg, C. B., Mach's Empirio-Pragmatism in Physical Science, thesis, Columbia University, New York, 1937.

A Symposium on Ernst Mach (ed. by J. Hintikka), *Synthese* **18** (1968) 132–301. This includes the following articles:

Robert S. Cohen, Ernst Mach: Physics, Perception and the Philosophy of Science. Also this volume, pp. 126–164.

Milič Čapek, Ernst Mach's Biological Theory of Knowledge. [Also in *Boston Studies in the Philosophy of Science*, vol. 4, pp. 400–420.]

S. G. Brush, Mach and Atomism.

A. Koslow, Mach's Concept of Mass: Program and Definition.

Wolfgang Yourgrau and Alwyn van der Merwe, Did Ernst Mach 'Miss the Target'?

M. Strauss, Einstein's Theories and the Critics of Newton. An Essay in Logico-Historical Analysis.

Joachim Thiele, Briefe deutscher Philosophen an Ernst Mach.

V. NON-ENGLISH PUBLICATIONS ON ERNST MACH SINCE 1930

Birk, Alfred, 'Zur Hundertjahrfeier zweier grossen Freunde. Ernst Mach – Josef Popper', *Ingenieur-Zeitschrift*, Teplitz Schönau **18** (1938), 12.

Blüh, Otto, 'Ernst Mach', *Brünner Tagesbote*, 15 February 1938.

Dugas, René, *La Théorie Physique au Sens de Boltzmann*. Part III, Chap. VI. 'Planck et Mach', p. 256. Griffon, Neuchâtel, 1959.

Heller, K. D., *Ernst Mach – Wegbereiter der Modernen Physik*, Springer Verlag, Vienna, 1964.

Herneck, Friedrich, 'Über eine Unveröffentliche Selbstbiographie Ernst Machs', *Wissenschaftliche Zeitschrift der Humboldt-Universität Berlin*, Mathem. naturwiss. Reihe, **6** (1956/57), 209.

Herneck, Friedrich, 'Ernst Mach, Eine bisher unveröffentlichte Autobiographie', *Phys. Bl.* **14** (1958), 385.

Herneck, Friedrich, 'Zu einem Brief Albert Einsteins an Ernst Mach', *Phys. Bl.* **15** (1959), 563.

Herneck, Friedrich, 'Wiener Physik vor 100 Jahren', *Phys. Bl.* **17** (1961), 455. N.B. Photographs of Mach and Boltzmann on p. 457.

Herneck, Friedrich, 'Zum Briefwechsel Albert Einsteins an Ernst Mach', *Forschungen und Fortschritte* **37** (1963), 239.

Herrmann, D. B., 'Ernst Mach und seine Stellung zur Doppler-Theorie', *Forschungen und Fortschritte* **40** (1966), 362.

Hönl, H., 'Ein Brief Albert Einsteins an Ernst Mach', *Phys. Bl.* **16** (1960), 571.

Kerkhof, F., 'Ernst-Mach Symposium am 11./12. März 1966 in Freiburg/Br.', *Phys. Bl.* **22** (1966), 223.

Löwy, Heinrich, 'Die Erkenntnistheorie von Popper-Lynkeus und ihre Beziehung zur Machschen Philosophie', *Die Naturwissenschaften* **20** (1932), 770.

Lohr, Ernst, 'Ernst Mach als Physiker', *Zeitschr. für die gesammte Naturwissenschaft* **4** (1938/39), 108.

Lübbe, Hermann, 'Positivismus und Phänomenologie', (Mach und Husserl) in: *Beiträge zur Philosophie und Wissenschaft*, Munich 1960, p. 161.

Mises, Richard von, 'Ernst Mach und die Empiristische Wissenschaftsauffassung. Zu Ernst Machs 100. Geburtstag am 18. Februar 1938', *Einheitswissenschaft* **7** (1938), 35. [Transl. in this volume, pp. 245–270.]

Nicolle, Jacques, 'Lénine, Mach et Paul Langevin', *La Pensée* **57** (1954), 66.

Oakland, Fridthjof, *Machs Elementlaere og Biologien*, Oslo 1947.

Sommerfeld, Arnold, 'Ernst Mach als Physiker, Psychologe und Philosoph', *Verhandl. Deutschen Physik. Gesellschaft*, Reihe 3, **19** (1938), 51.

Thiele, J., 'Bemerkungen zu einer Äusserung im Vorwort der *Optik* von Ernst Mach', *Schriftenreihe für Geschichte der Naturwiss., Technik und Medizin* **2** (1965), 10.

Thiele, J., 'William James und Ernst Mach', *Philosophia Naturalis* **9** (1966), 298.

Thiele, J., 'Briefe von Gustav Theodor Fechner und Ludwig Boltzmann an Ernst Mach', *Centaurus* **11** (1967), 222.

Thiele, J., 'Naturphilosophie und Monismus um 1900' (Briefe von W. Ostwald, Ernst Mach, Ernst Haeckel und Hans Driesch), *Philosophia Naturalis* **10** (1968), 295.

Thiele, J., 'Briefe Robert Lowies an Ernst Mach', *Isis* **59** (1968), 84.

Thiele, J., 'Ernst Mach und Heinrich Hertz, Zwei unveröffentlichte Briefe aus dem Jahre 1890', *Schriftenreihe für Geschichte der Naturwiss., Technik und Medizin* **5** (1968), 132.

Thiele, J., 'Ein zeitgenössisches Urteil über die Kontroverse zwischen Max Planck und Ernst Mach', *Centaurus* **13** (1968), 85.

'Symposium aus Anlass des 50. Todestages von Ernst Mach', veranstaltet vom Ernst-Mach-Institut, Freiburg/Br. am 11/12. März 1966 in Freiburg/Br. With contributions by G. v. Békésy, M. Bunge, F. v. Hayek, F. Herneck, D. B. Herrmann, N. Hiebert, H. Hönl, H. P. Jochim, R. E. Kutterer, E. Lesky, J. Mayerhoefer, W. F. Merzkirch, J. Pachner, F. Ratliff, N. Strauss, J. Thiele.

INDEX OF NAMES

SYNTHESE LIBRARY

Monographs on Epistemology, Logic, Methodology,
Philosophy of Science, Sociology of Science and of Knowledge, and on the
Mathematical Methods of Social and Behavioral Sciences

‡NICHOLAS RESCHER *et al.*, *Essays in Honor of Carl G. Hempel. A Tribute on the Occasion of his Sixty-Fifth Birthday*. 1969, VII + 272 pp. Dfl. 46.—

‡PATRICK SUPPES, *Studies in the Methodology and Foundations of Science. Selected Papers from 1951 to 1969*. 1969, XII + 473 pp. Dfl. 72.—

‡JAAKKO HINTIKKA, *Models for Modalities. Selected Essays*. 1969, IX + 220 pp. Dfl. 34.—

‡D. DAVIDSON and J. HINTIKKA: (eds.), *Words and Objections: Essays on the Work of W. V. Quine*. 1969, VIII + 366 pp. Dfl. 48.—

‡J. W. DAVIS, D. J. HOCKNEY, and W. K. WILSON (eds.), *Philosophical Logic*. 1969, VIII + 277 pp. Dfl. 45.—

‡ROBERT S. COHEN and MARX W. WARTOFSKY (eds.), *Boston Studies in the Philosophy of Science*. Volume V: *Proceedings of the Boston Colloquium for the Philosophy of Science 1966/1968*. 1969, VIII + 482 pp. Dfl. 58.—

‡ROBERT S. COHEN and MARX W. WARTOFSKY (eds.), *Boston Studies in the Philosophy of Science*. Volume IV: *Proceedings of the Boston Colloquium for the Philosophy of Science 1966/1968*. 1969, VIII + 537 pp. Dfl. 69.—

‡NICHOLAS RESCHER, *Topics in Philosophical Logic*. 1968, XIV + 347 pp. Dfl. 62.—

‡GÜNTHER PATZIG, *Aristotle's Theory of the Syllogism. A Logical-Philological Study of Book A of the Prior Analytics*. 1968, XVII + 215 pp. Dfl. 45.—

‡C. D. BROAD, *Induction, Probability, and Causation. Selected Papers*. 1968, XI + 296 pp. Dfl. 48.—

‡ROBERT S. COHEN and MARX W. WARTOFSKY (eds.), *Boston Studies in the Philosophy of Science*. Volume III: *Proceedings of the Boston Colloquium for the Philosophy of Science 1964/1966*. 1967, XLIX + 489 pp. Dfl. 65.—

p.t.o.